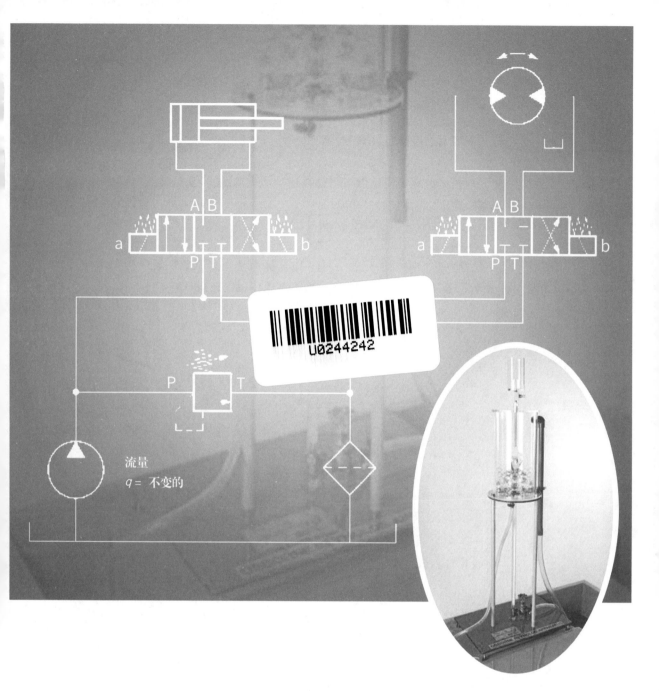

FLUID MECHANICS AND FLUID POWER LABORATORY MANUAL
流体力学及液压传动实验教程

主编 ◎ 张宏

大连理工大学出版社
Dalian University of Technology Press

图书在版编目(CIP)数据

流体力学及液压传动实验教程 / 张宏主编. --大连：大连理工大学出版社，2021.2
ISBN 978-7-5685-2628-9

Ⅰ. ①流… Ⅱ. ①张… Ⅲ. ①流体力学－高等学校－教材②液压传动－高等学校－教材 Ⅳ. ①O35②TH137

中国版本图书馆CIP数据核字(2020)第135482号

流体力学及液压传动实验教程
LIUTI LIXUE JI YEYA CHUANDONG SHIYAN JIAOCHENG

大连理工大学出版社出版
地址：大连市软件园路80号 邮政编码：116023
发行：0411-84708842 邮购：0411-84708943 传真：0411-84701466
E-mail：dutp@dutp.cn URL：http://dutp.dlut.edu.cn
大连图腾彩色印刷有限公司印刷 大连理工大学出版社发行

幅面尺寸：185mm×260mm　　印张：10.75　　字数：241千字
2021年2月第1版　　　　　　　　　　　2021年2月第1次印刷

责任编辑：李宏艳　　　　　　　　　　　　　　责任校对：周　欢
封面设计：奇景创意

ISBN 978-7-5685-2628-9　　　　　　　　　　　　　定　价：25.00元

本书如有印装质量问题，请与我社发行部联系更换。

Preface

Introduction

This book is concerned with the Fluid Mechanics and Hydraulic experiments that future engineers should master. For the mechanical engineering related majors, it is crucial that the Fluid Mechanics fundamentals and its applications in the field of Hydraulic in various forms should be thoroughly understood.

Fluid Mechanics is defined as the science that deals with the behavior of fluid at rest (fluid statics) or in motion (fluid dynamics), and the interaction of fluids with solids or other fluids at the boundaries. Fluid Mechanics is also referred to as fluid dynamics by considering fluids at rest as a special case of motion with zero velocity. Also, fluid mechanics itself is divided into several categories. The study of the motion of fluids that can be approximated as incompressible (such as liquids, especially water, and gases at low speeds) is usually referred to as hydrodynamics. A subcategory of hydrodynamics is hydraulics, which deals with liquid flows in pipes and open channels. Of course it also has categories of Gas dynamics and aerodynamics.

According to the latest requirements of mechanical major, great efforts are given in the Fluid Mechanics and Hydraulic parts in this book. Consequently, the contents in this book will be contributed by two parts: Fluid Mechanics (Lab 1 to Lab 3) and hydraulic (Lab 4 to Lab 14). The fundamental theories will be presented, ways of operation methods in order to shed light on theories will be shown, and illuminating lab reports will be developed. The experiment equipment of Fluid Mechanics is based on Armfield which is made in English, and the hydraulic experiments are equipped by Bosch Rexroth from Germany. Every experiment requires at least 2 hours and they are independent. It can be selected according to diverse requirements. It not only encourages the undergraduate student to enhance the operation ability, but also can be extended to treat advanced topics in graduate-level courses and to deal with realistic problems in an industrial context.

Note: Units are not uniform in this book, which take the common usage in engineering applications.

Acknowledgments

As in the case of this edition, I am indebted to numerous Fluid Power and Fluid Mechanics equipment-manufacturing companies for permitting the inclusion of their photographs and other illustrations in this book.

I especially thank professor Jan Lugowski and professor Steven Widmer in Purdue University for providing many helpful suggestions and comments for improving this edition. I also thank Chengxiang Wang from Bosh Rexroth for reviewing this edition and providing many helpful suggestions. Thanks to the Press of Dalian University of Technology for handling the production of this edition. I also thank laboratory technician Liu Jianwei for debugging and my students Tang Ming (debugged Lab 11), Ding Lan (debugged Lab 14, Lab 15 and Lab 16), Zhang Lu and Li Chenchen for the contribution to the materials organizing.

Zhang Hong
2021.1

Contents

Lab 1　Reynolds Experiment 1
 Objectives 1
 Equipment Preparation 1
 Principle 1
 Equipment Description 3
 Part 1　Observing Flow Status 4
 Part 2　Determining Reynolds Number 5

Lab 2　Bernoulli's Experiment 6
 Objectives 6
 Equipment Preparation 6
 Discussion 8
 Procedure 9

Lab 3　Fluid Friction 11
 Objectives 11
 Equipment Preparation 11
 Part 1　Fluid Friction in a Smooth Bore Pipe 13
 Part 2　Head Loss Due to Pipe Fittings 15
 Part 3　Fluid Friction in a Roughened Pipe 17
 Part 4　Fluid Friction in an Orifice Plate or Venturi 18

Lab 4　Fluid Power Components and Circuits 21
 Objectives 21
 Equipment Preparation 21
 Software Introduction 21
 Procedure 23
 Part 1　Recognize the Components on the Workstation 23
 Part 2　Familiar with the Symbol of the Components 24
 Part 3　Recognize the Circuits 24

Lab 5　Pump Disassembly and Performance 26
 Objectives 26
 Equipment Preparation 26

| Part 1 | Pump Disassembly | 26 |
| Part 2 | Pump Performance | 29 |

Lab 6 Cylinder Circuit Operation 38
Objectives 38
Equipment preparation 38
Discussion 38
Part 1 Normal Cylinder Operation 39
Part 2 Regenerative Cylinder Operation 41

Lab 7 Hydraulic Motor 44
Objectives 44
Equipment Preparation 44
Discussion 44
Procedure 45

Lab 8 Valve Disassembly 48
Objectives 48
Discussion 48
Software Introduction 48

Lab 9 Analysis of Flow Control Valves 50
Objectives 50
Equipment Preparation 50
Discussion 50
Part 1 Throttle Valve 51
Part 2 Speed-regulating Valve 56

Lab 10 Familiar with Hydraulic Circuits 59
Objectives 59

Lab 11 Proportional Control System 60
Objectives 60
Equipment Preparation 60
Discussion 60
Part 1 Working Principle of Proportional Hydraulic Valve 60
Part 2 Characteristic Curve of the Proportional Direction Valve 65
Part 3 PID Control Simulation and Experiment 67

Contents

Lab 12 Hydraulic Throttle Control Experiment for Engineering Machinery ········ 76
    Objectives ·· 76
    Equipment Preparation ·· 76
    Principle ··· 77
    Preparation ··· 83
    Procedure ··· 86

Lab 13 Hydraulic LS Control Experiment for Engineering Machinery ············· 88
    Objectives ·· 88
    Equipment Preparation ·· 88
    Principle ··· 88
    Preparation ··· 97
    Procedure ··· 100

Lab 14 LUDV Control Experiment for Engineering Machinery ······················ 102
    Objectives ·· 102
    Equipment Preparation ·· 102
    Principle ··· 102
    Preparation ··· 110
    Solution to the Cavitation Phenomenon ·· 110
    Procedure ··· 113

Lab Report ··· 115
    Lab Report 1   Reynolds Experiment ··· 117
    Lab Report 2   Bernoulli's Experiment ··· 119
    Lab Report 3   Fluid Friction ··· 121
    Lab Report 4   Fluid Power Components and Circuits ··································· 127
    Lab Report 5   Pump Disassembly and Performance ····································· 131
    Lab Report 6   Cylinder Circuit Operation ··· 137
    Lab Report 7   Hydraulic Motor ·· 141
    Lab Report 8   Valve Disassembly ·· 145
    Lab Report 9   Analysis of Flow Control Valves ·· 149
    Lab Report 10 Familiar with Hydraulic Circuits ·· 155
    Lab Report 11 Proportional Control System ·· 157
    Lab Report 12 Hydraulic Throttle Control Experiment for Engineering
                          Machinery ··· 159
    Lab Report 13 Hydraulic Load Sensing Control Experiment for Engineering
                          Machinery ··· 161
    Lab Report 14 LUDV Control Experiment for Engineering Machinery ········ 163

Lab 1　Reynolds Experiment

Objectives

(1) Observe the state of laminar, transitional and turbulent flow in a test pipe and their switching processes;

(2) Determine the critical Reynolds number Re and master the criterion of flow states in pipeline.

Equipment Preparation

The experiment makes use of F5 from Armfield of England (as shown in Figure 1.2).

Principle

In 1883, Osborne Reynolds (British physicist 1842—1912) used the experimental device and found there exist two kinds of flow states, laminar flow and turbulence flow. Reynolds contributed a lot and not only found out the above flow states, but also drawn out the criterion to judge them. The criterion Re is a parameter of dimensionless one, which can be used for any Newtonian fluid flow state transform in pipeline of any diameter.

Laminar flow represents a steady flow condition where all streamlines follow parallel paths, there being no interaction (mixing) between shear planes. The observer will find that the dye (indicator) will remain as a solid and easily identifiable component of flow under this condition.

Turbulent flow represents an unsteady flow condition where streamlines interaction causes collapsing of shear planes and mixing of fluid. The observer will find that the dye will become dispersed in mixing fluid and will not remain as a unit component of flow under this condition.

The criterion of Re depends on three parameters (as shown in Table 1.1 and Figure 1.1):

(1) mean velocity v of fluid;
(2) pipe diameter D;
(3) kinematic viscosity of fluid ν.

Table 1.1 Nomenclature in This Project

Name	Symbol	Unit
Reynolds Number	Re	—
Friction Factor	λ	—
Kinematic Viscosity	ν	m^2/s
Pipe Diameter	D	m
Mean Velocity	v	m/s
Higher Critical Velocity	v^{crit}	m/s
Lower Critical Velocity	v_{crit}	m/s
Flow Rate	q	m^3/s
Shear Stress at the Pipe Wall	τ	N/m^2

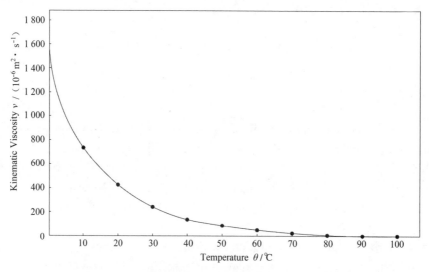

Figure 1.1 Viscosity Curve

$$Re = \frac{\text{mean velocity} \times \text{pipe diameter}}{\text{kinematic viscosity}} = \frac{v \cdot D}{\nu} = \frac{4q}{\pi D \nu}$$

Re is independent of pressure. If Re, which represents the flow status of a fluid in a pipe line, is less than 2 000, the flow will be laminar. If Re is greater than 2 800, the flow will be turbulent. If Re lies between these values, the flow may be either laminar or turbulent. Thus, there is no fixed value for the critical velocity v_{crit}, which is dependent on factors such as pipe internal surface condition and kinematic viscosity of the fluid.

Because fluid flows between the fixed boundaries of the pipe line, the greatest resistance in it will be found at the bounding surfaces. This will cause a retardation of the fluid particle velocity, which will promote a "velocity profile" in relation to the free steam velocity of the fluid particles on the central axis of the flow. This typical paraboloid curve will be observed through the apparatus under laminar flow conditions.

Lab 1 Reynolds Experiment

Notes: (1) In practice, the inner diameter of the sight tube shall be measured before it is assembled into the equipment. It is 13 mm in this project.

(2) Friction factor shall be calculated using the following equations:

Laminar regime (Poiseuille) $\quad \lambda = \dfrac{64}{Re}$

Turbulent regime (Blasius) $\quad \lambda = 0.316 Re^{0.25}$

Equipment Description

The equipment is shown as Figure 1.2.

1—Dye Reservoir; 2—Dye Injector; 3—Cover Plane; 4—Stilling Tank; 5—Bellmouth;
6—Dye Injector Tube; 7—Tank Stand; 8—Out Flow

Figure 1.2 Osborne Reynolds Apparatus

The apparatus is mounted on a bench and designed for the vertical fluid flow through a precision bore in the glass tube. Using vertical direction fluid flow can compensate for the effects of any small deviations between the density of dye and the working fluid. In the horizontal direction, the density of dye could not be precisely equal to that of the fluid, which would lead to deviation of the trace in vertical axis.

The working fluid can be supplied from any supply point (small bores) by means of the flexible hose. It enters a cylindrical constant head tank through a ring diffuser and a particle bed in order to eliminate any gross variations of the fluid velocity.

Therefore, this tank provides uniform, low velocity head conditions for upstream of the entry to the vertically mounted pipe test section. Fluid enters this section through a profiled bell mouth, designed to accelerate uniformly the fluid without any spurious inertial effects.

The cylindrical pipe test section is mounted inside a fabricated shroud which provides uninterrupted white background for the observations of the dye trace behavior.

Dye solution flows into the test section through a fine diameter stainless steel tube, and flow-rate of the dye is controlled by a valve on the outlet of the reservoir. The dye injection system can be removed easily for cleaning and maintenance.

The flow rate of working fluid through the test section is regulated by a needle globe valve located in the base of the apparatus. The flow rate may be measured either by volume or by weighing, and the result is independent of the kinematic viscosity of the fluid, which can be changed in different fluids or fluid temperature.

The whole test equipment is rigidly mounted on a solid fabricated steel support, and a device is used for leveling the apparatus in order to keep the test section in the vertical direction.

This project consists of two separate tasks. Part 1 is an exercise to observe laminar, transitional and turbulent flow in a test pipe. Part 2 is determining Reynolds Number Re on the critical velocities.

Part 1 Observing Flow Status

Lower the dye injector until it is just above the inlet of bell mouth. Open the inlet valve and the water will enter the stilling tank. Ensure a small overflow relieves from the upper drain outlet so that the water level can keep constant.

Then we open the control valve a little, and the dye will go into the pipe in very low flow rate. In this case, the dye is drawn out through the center of the pipe. Increasing the flow rate will produce eddies in the dye until it completely diffuses into the water.

The dye experiment demonstrates there is a critical velocity of the water when it enters the tube in reasonably steady state, and it also shows another critical velocity of the water in a highly unstable state. In other words, two critical velocities exist in different situations, one is steady flow which is changed into eddies, and the other is eddies which are changed into turbulence as Figure 1.3 shows.

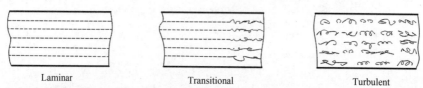

Laminar   Transitional   Turbulent

Figure 1.3 Flow in a Pipe

Lab 1 Reynolds Experiment

To observe the velocity profile, open the dye reservoir needle valve and deposit a drop of dye in the pipe. Open the flow control valve and observe that the drop presents a three dimensional parabolic profile.

Part 2 Determining Reynolds Number

Lower the dye injector until it is just above the inlet of the bell mouth. Open the inlet valve and the water will enter the stilling tank. Ensure a small overflow relieves from the upper drain outlet so that the water level can keep constant.

Allow the water to settle for five minutes and measure the water temperature using a thermometer during this time.

Open the flow control valve a little and adjust the dye control needle valve to get a slow flow of the dye injection. Measure the flow rate when the laminar flow changes into eddies. Also record the flow rate at which the dye becomes fully mixed with the water and appears to be the turbulent flow.

Reverse the operation and observe critical points of the flow state.

Repeat the experiment and fill the data in Table R.1.1 of lab report.

Lab 2 Bernoulli's Experiment

Objectives

(1) Investigate the validity of the Bernoulli equation when it is applied to the steady flow of water in a duct;

(2) Learn the inter-conversion among various mechanical energy, while the total energy remains constant in steady, incompressible flow with negligible friction.

Equipment Preparation

In order to carry out the demonstration of the Bernoulli's theorem verification, we need some equipments as follows:

(1) The F1-10 hydraulics workbench from Armfield of England is shown in Figure 2.1, which we can measure the flow according to timed volume collection;

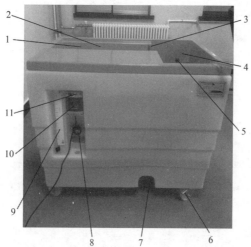

1—Slots for Stilling Baffle; 2—Flow Channel; 3—Weir Carrier; 4—Volumetric Tank; 5—Dump Valve Actuator; 6—Castor; 7—Sump Tank Drain Valve; 8—Flow Control Valve; 9—Sight Gauge and Level Scale; 10—RCD; 11—Power Switch

Figure 2.1 F1-10 Hydraulic Workbench

(2) The F1-15 Bernoulli's apparatus test equipment as Figure 2.2 shows. The position datum and corresponding assumed datum as Figure 2.3 and Table 2.1 show. It is supplied by Armfield from England;

(3) A stopwatch for timing in the flow measurement.

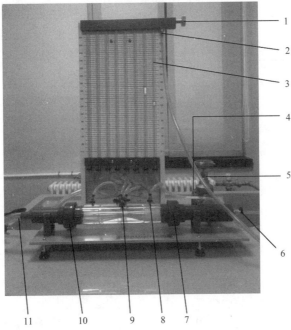

1—Air Bleed Screw; 2—Air Bleed Connection; 3—Manometer Tubes; 4—Water Outlet;
5—Flow Control Valve; 6—Total Head Tube Entry Gland; 7—Union;
8—Total Head Tube(Inside Test Section); 9—Total Section(Reversible);
10—Union; 11—Water Inlet

Figure 2.2 F1-15 Bernoulli Apparatus Structure

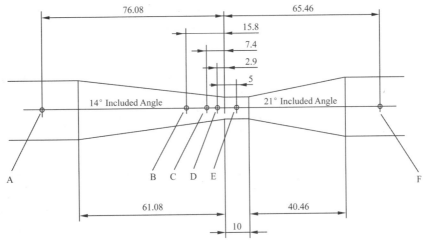

Figure 2.3 Position Datum in Figure 2.2

Table 2.1 Assumed Datum Position

Tapping Position	Manometer Legend	Diameter/mm	Tapping Position	Manometer Legend	Diameter/mm
A	h_1	25.0	D	h_4	10.7
B	h_2	13.9	E	h_5	10.0
C	h_3	11.8	F	h_6	25.0

Note: The assumed datum position is at tapping A associated with h_1.

Discussion

Method: Use a rigid convergent/divergent tube of known geometry to measure flow rate and both static and total pressure head for a range of steady flow rates.

The Bernoulli equation represents the conservation of mechanical energy for a steady, incompressible, frictionless flow:

$$\frac{p_1}{\rho g}+\frac{v_1^2}{2g}+z_1=\frac{p_2}{\rho g}+\frac{v_2^2}{2g}+z_2 \qquad (2\text{-}1)$$

where p is static pressure tested at a side hole; v is fluid velocity; z is vertical elevation of the fluid; $z_1=z_2$ is for a horizontal tube.

The equation can be derived from the Euler equations by integration. It also can be derived from energy conservation principles. Derivation of the Bernoulli equation is beyond the scope of this theory. If the tube is horizontal, the difference in height will not be in consideration.

$$z_1=z_2 \qquad (2\text{-}2)$$

$$\frac{p_1}{\rho g}+\frac{v_1^2}{2g}=\frac{p_2}{\rho g}+\frac{v_2^2}{2g} \qquad (2\text{-}3)$$

Hence, with Bernoulli's apparatus, the static pressure head p can be measured using a manometer directly from a side hole with pressure tapping.

The manometer actually measures the static pressure head h in meters, which is related to the pressure p in the following:

$$h=\frac{p}{\rho g} \qquad (2\text{-}4)$$

So the Bernoulli equation can be written in a revised form:

$$h_1+\frac{v_1^2}{2g}=h_2+\frac{v_2^2}{2g} \qquad (2\text{-}5)$$

The velocity-related part of the total pressure head is called the dynamic pressure head.

The total pressure head h^0 can be measured from a probe with an end hole facing into the flow, which can bring the flow to rest locally at the probe end. Thus:

$$h^0=h+\frac{v^2}{2g} \qquad (2\text{-}6)$$

According to the Bernoulli equation, it shows that:

$$h_1^0=h_2^0 \qquad (2\text{-}7)$$

The velocity of the flow can be measured by testing the volume V of the flow over a time period T. This gives the rate of volume flow as $q_V = \dfrac{V}{T}$, which in turn gives the velocity of flow through a defined area A:

$$v = \frac{q_V}{A} \tag{2-8}$$

For an incompressible fluid, conservation of mass means that the flow rate will be also conserved. That is:

$$A_1 v_1 = A_2 v_2 \tag{2-9}$$

Procedure

1. Setting up Equipment

Set up the Bernoulli apparatus F1-15 on the hydraulic bench F1-10 so that its base is in horizontal position, and this is necessary for accurate height measurement by the manometers.

2. Set the Direction of the Test Section

Ensure that the 14° tapered part of the test-section is converging in the direction of flow. If you need to reverse the test-section, the total pressure head probe must be withdrawn before releasing the mounting couplings.

3. Connect the Water Inlet and Outlet

Ensure that the rig outflow tube is installed above the volumetric tank, in order to facilitate timed volume collections. Connect the rig inlet to the bench flow supply, close the bench valve and the apparatus flow control valve, and start the pump. Gradually open the bench valve to fill the test rig with water.

4. Bleeding the Manometer

In order to bleed air from pressure tapping points and manometers, we need close both the bench valve and the rig flow control valve, open the air bleed screw and remove the cap from the adjacent air valve. Connect a small bore tube between the air valve and the volumetric tank. Now, open the bench valve and allow the flow to go through the manometers to push all of the air out, and then tighten the air bleed screw and partly open the bench valve and test rig flow control valve. Next, open the air bleed screw slightly to allow air to enter the top of the manometers (you may need to adjust both valves), and retighten the screw when the level of manometer reaches a convenient height. The maximum volume flow rate will be determined by the maximum h_1 and minimum h_5 manometer readings both on scale.

If required, the manometer levels can be adjusted further by using the air bleed screw and the hand pump supplied. The air bleed screw controls the air flow of the air valve, which should be open when using the hand pump. To retain the pressure in the system, the screw must be closed after pumping.

5. Recording a Set of Results

Reading should be taken at three flow rates. Finally, you may reverse the test section in order to see the effects of a more rapid converging section. Fill the data into lab report Table R.2.1 and answer the corresponding questions in lab report.

6. Setting the Flow Rate

Take the first set of readings at the maximum flow rate, and then reduce the volume flow rate to give the h_1-h_5 head difference of about 50 mm. Finally repeat the whole process for one further flow rate, and set to give the h_1-h_5 difference approximately half way between that obtained in the above two tests.

7. Reading the Static Head

Take readings of the h_1-h_5 manometers when the levels have steadied. Ensure that the total pressure probe is retracted from the test-section.

8. Timed Volume Collection

You should carry out a timed volume collection, using the volumetric tank F1-10 to determine the volume flow rate. This is achieved by closing the ball valve and measuring (with a stopwatch) the time taken to accumulate a known volume of fluid in the tank, which is read from the sight glass. You should collect fluid for at least one minute to minimize timing errors. Again the total pressure probe should be retracted from the test-section during these measurements.

9. Reading the Total Pressure Head Distribution

Measure the total pressure head distribution by traversing the total pressure probe along the length of the test section. The datum line is the side hole pressure tapping associated with the manometer h_1. A suitable starting point is 1 cm upstream of the beginning of the 14° tapered section and measurements should be made at 1 cm intervals along the test-section length until the end of the divergent(21°) section.

10. Reversing the Test Section

Ensure that the total pressure probe is fully withdrawn from the test-section(but not pulled out of its guide in the downstream coupling). Unscrew the two couplings, remove the test-section and reverse it, and then reassemble by tightening the couplings.

Lab 3 Fluid Friction

Objectives

(1) etermine the relationship between head loss due to fluid friction and velocity for flow of water through smooth bore pipes and confirm the head loss predicted by a pipe friction equation;

(2) Determine the head loss associated with the flow of water through standard fittings used in plumbing installations;

(3) Determine the relationship between fluid friction coefficient and Reynolds number for the flow of water through a pipe having a roughened bore;

(4) Demonstrate the application of differential head devices in the measurement of flow rate and velocity of water in a pipe.

Equipment Preparation

The experiment makes use of C6-MK Ⅱ-10 from Armfield of England.

The Armfield C6-MK Ⅱ-10 fluid friction apparatus (as Figure 3.1 shows) is designed to allow the detailed study of the fluid friction head losses which occurs when an incompressible fluid flows through pipes, bends, valves and pipe flow metering devices.

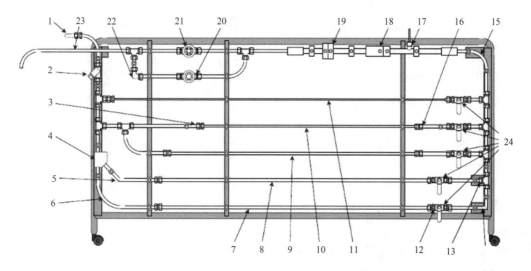

Figure 3.1 General Arrangement of C6-MK Ⅱ-10 Fluid Friction Apparatus

The test pipes and fittings are mounted on a tubular frame carried with castors. Water is fed in from the hydraulic bench via the barbed connector 1, flows through the network of pipes and fittings, and is fed back into the volumetric tank via the exit tube 23.

The pipes are arranged as Figure 3.2 shows to provide facilities for testing as follows:

(1) an in-line strainer 2;

(2) an artificially roughened pipe 7;

(3) smooth bore pipes of 4 different diameters 8, 9, 10 and 11;

(4) a long radius 90° bend 6;

(5) a short radius 90° bend 15;

(6) a 45° "Y" 4;

(7) a 45° elbow 5;

(8) a 90° "T" 13;

(9) a 90° miter 14;

(10) a 90° elbow 22;

(11) a sudden contraction 3;

(12) a sudden enlargement 16;

(13) a pipe section made of clear acrylic with a Pitot static tube 17;

(14) a Venturi made of clear acrylic 18;

(15) an orifice meter made of clear acrylic 19;

(16) a ball valve 12;

(17) a globe valve 20;

(18) a gate valve 21.

Test pipe diameters are as follows:

(1) 19.1 mm × 17.2 mm;

(2) 12.7 mm × 10.9 mm;

(3) 9.5 mm × 7.7 mm;

(4) 6.4 mm × 4.5 mm;

(5) 19.1 mm × 15.2 mm (artificially roughened).

Distance between tapings is 1.00 m.

Lab 3 Fluid Friction

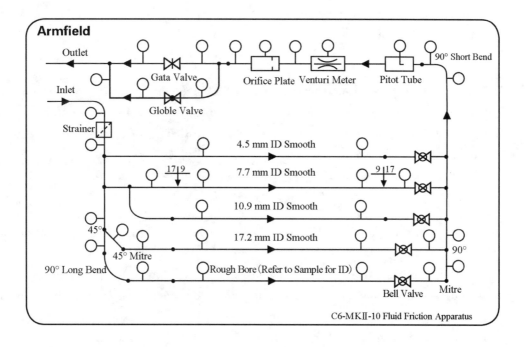

Figure 3.2 Schematic Diagram

Part 1 Fluid Friction in a Smooth Bore Pipe

Theory

Osborne Reynolds demonstrated that two types of flow may exist in a pipe:

(1) laminar flow at lower velocities where $h \propto u$;

(2) turbulent flow at higher velocities where $h \propto u^n$.

Where h is the head loss due to friction and u is the fluid velocity. These two types of flow are separated by a transition phase where no definite relationship between h and u exists.

Graphs of h versus u and $\lg h$ versus $\lg u$ are shown as Figure 3.3.

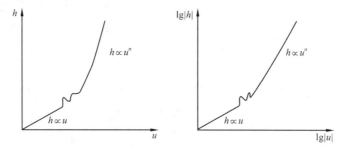

Figure 3.3 Graphs of h versus u and $\lg h$ versus $\lg u$

Furthermore, for a circular pipe flowing full, the head loss due to friction may be calculated from the equation:

$$h = \frac{4fLu^2}{2gd} \quad \text{or} \quad \lambda \frac{L}{d} \cdot \frac{u^2}{2g} \tag{3-1}$$

where L is the length of the pipe between tapings; d is the internal diameter of the pipe; u is the mean velocity of water through the pipe in m/s; g is the acceleration due to gravity in m/s² and f is pipe friction coefficient. Note that the American equivalent of the British term f is λ, where $\lambda = 4f$.

The Reynolds number Re can be found using the following equation:

$$Re = \frac{\rho u d}{\mu} \tag{3-2}$$

where μ is the dynamic viscosity (1.15×10^{-3} N·s/m² at 15 ℃) and ρ is the density (999 kg/m³ at 15 ℃).

Having established the value of Reynolds number for flow in the pipe, the value of f may be determined using a Moody diagram, and a simplified version of which is shown in Figure 3.4.

Equation (3-1) can be used to determine the theoretical head loss.

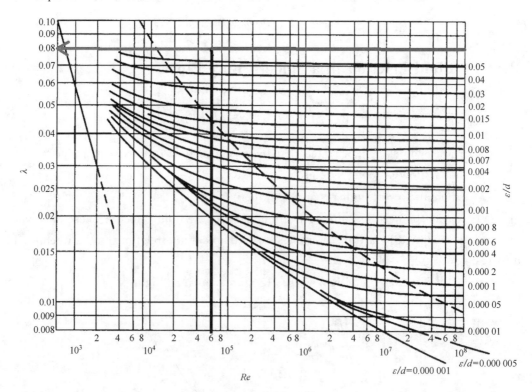

Figure 3.4 Moody Diagram

Lab 3 Fluid Friction

Procedure

Prime the pipe network with water. Open and close the appropriate valves to obtain flow of water through the required test pipe.

Take readings at several different flow rates, altering the flow using the control valve on the hydraulics bench (ten readings are sufficient to produce a good head-flow curve).

Measure flow rates using the volumetric tank (if using C6-MK II software, flow rate is measured directly). For small flow rates use the measuring cylinder. Measure head loss between the tappings using the portable pressure meter or pressurized water manometer as appropriate.

Obtain readings on all four smooth test pipes.

Measure the internal diameter of each test pipe sample using a vernier calliper.

Results

All readings should be tabulated and filled in Table R. 3. 1 of the lab report.

Plot a graph of h versus u for each size of pipe in Figure R. 3. 1. Identify the laminar, transition and turbulent zones on the graphs.

Confirm that the graph is a straight line for the zone of laminar flow $h \propto u$.

Plot a graph of lg h versus lg u for each size of pipe in Figure R. 3. 1. Confirm that the graph is a straight line for the zone of turbulent flow $h \propto u^n$. Determine the slope of the straight line to find n.

Estimate the value of Reynolds number $\left(Re = \dfrac{\rho u d}{\mu}\right)$ at starting and finishing the transition phase. These two values of Re are called the upper and lower critical velocities.

Confirm that the head loss can be predicted using the pipe friction equation provided the velocity of the fluid and the pipe demensions are known.

It is assumed that the dynamic viscosity μ is 1.15×10^{-3} N·s/m^2 at 15 ℃ and the density ρ is 999 kg/m^3 at 15℃.

Part 2 Head Loss Due to Pipe Fittings

Theory

Head loss in a pipe fitting is proportional to the velocity head of the fluid flowing through the fitting:

$$h = \frac{Ku^2}{2g} \tag{3-3}$$

where K is the fitting "loss factor"; u is the mean velocity of water through the pipe in m/s; g is the acceleration due to gravity in m/s².

As velocity changes in the contraction and in the enlargement, it is necessary to correct the measured loss to account for the change in velocity head.

Note: A flow control valve is a pipe fitting which has an adjustable "K" factor. The minimum valve of "K" and the relationship between stem movement and "K" factor are important in selecting a valve for an application.

Equipment Setup

Measure the differential head between tapings on fittings and test valves.

There is an additional equipment required: stop watch.

The following fittings and valves are available for test as Table 3.1 shows(numbers in brackets refer to Figure 3.1 in the equipment diagrams).

Table 3.1 Available Fittings and Valves

Valves		Valves	
Sudden Contraction	3	Globe Valve	20
Sudden Enlargement	16	In Line Strainer	2
Ball Valve	12	90° Elbow	22
45° Elbow	22	90° Short Radius Bend	15
45° Mitre	5	90° Long Radius Bend	6
45° Y Junction	4	90° T Junction	13
Gate Valve	21		

If using the C6-50 data logging accessory, ensure that the platform powered and connected to the PC via the USB connection. Load the C6-MK II software and choose exercise B.

Procedure

Prime the network with water. Open and close the appropriate valves to obtain flow of water through the required fittings.

Take readings at several different flow rates, altering the flow using the control valve on the hydraulics bench.

Measure differential head between tappings on each fitting using the hand held pressure meter, sensors or pressurized water manometer.

Lab 3　Fluid Friction

Results

All readings should be tabulated in Table R. 3. 2 of the lab report.

Confirm that K is a constant for each fitting over the range of test flow rates.

Plot a graph of K factor against valve opening for each test valve in Figure R. 3. 2. Note the differences in characteristics.

Note: Measured head loss across the contraction and the enlargement must be corrected for the change in velocity head (due to the change in pipe diameter) to obtain the true head loss K factor.

Part 3　Fluid Friction in a Roughened Pipe

Theory

The head loss due to friction in a pipe is given by:

$$h = \frac{4fLu^2}{2gd} \quad \text{or} \quad \frac{\lambda L u^2}{2gd} \tag{3-4}$$

where L is the length of the pipe between tappings; d is the internal diameter of the pipe; u is the mean velocity of water through the pipe in m/s; g is the acceleration due to gravity in m/s^2; f is the pipe friction coefficient. Note that the American equivalent of the British term f is λ, where $\lambda = 4f$.

The Reynolds number can be found using the following equation:

$$Re = \frac{\rho u d}{\mu} \tag{3-5}$$

where μ is the dynamic viscosity (1.15×10^{-3} N · s/m^2 at 15 ℃); ρ is the density (999 kg/m^3 at 15 ℃).

Having established the value of Reynolds number for flow in the pipe, the valve may be determined using a Moody diagram, a simplified version of which is shown as Figure 3. 4.

Equation (3-4) can be used to determine the theoretical head loss.

Equipment Setup

To obtain a series of readings of head loss at different flow rates through the roughened test pipes, there are additional equipments required: a stop watch and an internal Vernier calliper.

Open and close the ball valve as required to obtain the flow through only the roughened pipe.

If using the C6-50 data logging accessory, ensure that the console is powered and connect to the PC via the USB connection. Load the C6-MK II software and choose exercise C.

Procedure

Prime the pipe network with water. Open and close the appropriate valves to obtain flow of water through the roughened pipe.

Take readings at several different flow rates, altering the flow using the control valve on the hydraulics bench (ten readings are sufficient to produce a good head-flow curve).

Measure flow rates using the volumetric tank (if using C6-MK II software, flow rate is measured directly). For small flow rates use the measuring cylinder.

Measure head loss between the tappings using the hand-hold meter, sensors or manometer as appropriate.

Estimate the nominal internal diameter of the test pipe sample using a Vernier calliper(not supplied). Estimate the roughness factor ε/d.

Results

Filling the results in Table R. 3. 3 of the lab report and corresponding blanks. Draw the graph of pipe friction coefficient versus Reynolds number in Figure R. 3. 3.

Part 4 Fluid Friction in an Orifice Plate or Venturi

Theory

For an orifice plate or Venturi, the flow rate and differential head are related by Bernoulli equation with a discharge coefficient added to account for losses:

$$q = C_d \cdot A_0 \sqrt{\frac{2g \cdot \Delta h}{1-(A_0/A_1)^2}} \quad (3\text{-}6)$$

where q is the flow rate in m³/s; C_d is the discharge coefficient ($C_d = 0.98$ for a Venturi, 0.62 for an orifice plate); A_0 is the area of the throat or orifice in m² ($d_0 = 14$ mm for the Venturi, 20 mm for the orifice plate); A_1 is the area of the pipe upstream m² ($d_1 = 24$ mm); Δh is the differential head in metres of water; g is the acceleration due to gravity in m/s².

For a Pitot tube, the differential head measured between the total and static tappings is equivalent to the velocity head of the fluid:

$$\frac{u^2}{2g} = h_1 - h_2 \quad (3\text{-}7)$$

$$u = \sqrt{2g(h_1 - h_2)} \quad (3\text{-}8)$$

Lab 3 Fluid Friction

where u is the mean velocity of water through the pipe in m/s; $h_1 - h_2$ is the differential head in metres of water; g is the acceleration due to gravity in m/s^2.

Equipment Setup

We will obtain a series of readings of head loss at different flow rates through an orifice plate, a Venturi meter and a Pitot tube.

There is an additional equipment required: stopwatch.

Open all ball valves to achieve the minimum restriction to flow.

If using the C6-50 data logging accessory, ensure that the console is powered and connected to the PC via the USB connection. Load the C60-MK II software and choose exercise D.

Procedure (Using the Venturi and Orifice Plate)

Prime the pipe network with water. Open the appropriate valves to obtain flow of water through the flowmeter.

Obtain readings from the Venturi and orifice plate at different flow rates from minimum to maximum flow, altering the flow rate using the control valve on the hydraulic bench. At each setting measure the differential head produced by each flowmeter, the head loss across each flowmeter and the corresponding volume flow rate.

Note: To measure the differential head developed by the orifice plate or Venturi (for the purpose of flow measurement) connect the probes from the appropriate manometer to the two tappings on the flowmeter body, upstream and at the throat (do not use the downstream tapping in the pipe). To measure the head loss across the orifice plate or Venturi connect the probes from the water manometer to the upstream tapping on the flowmeter body and the tapping in the pipe downstream of the device (do not use the throat tapping).

Results (for the Venturi and Orifice Plate)

All readings should be tabulated in Table R. 3. 4 and draw the graph of pipe friction coefficient versus Reynolds number in Figure R. 3. 4.

Procedure (for the Pitot Tube)

Ensure that the nose of the Pitot tube is directly facing the direction of flow and located on the centre line of the pipe.

Obtain readings from the Pitot tube at different flow rates from minimum to maximum flow. At each setting of the flow control valve measure the differential head produced by the Pitot tube and the corresponding volume flow rate.

At the maximum flow setting unscrew the sealing gland sufficiently to allow the

Pitot tube to move. Traverse the tube across the diameter of the pipe and observe the change in differential head. Estimate the average reading obtained and compare this with the maximum reading at the centre of the pipe.

Results (for the Pitot Tube)

All readings should be tabulated in Table R. 3. 5 and draw the graph of pipe friction coefficient versus Reynolds number in Figure R. 3. 5.

Lab 4   Fluid Power Components and Circuits

Objectives

Be familiar with various fluid power components such as valves, conductors, actuators, pumps, and fittings. By completing this exercise, the student should be able to:
(1) Identify the typical components on the laboratory workstations;
(2) Understand the component interfaces and operation of workbench;
(3) Become familiar with typical fluid power symbols;
(4) Do practice with Automation Studio software;
(5) Draw a standard cylinder and hydraulic motor circuit.

Equipment Preparation

The training device DS4 workbench supplied by Bosch Rexroth from Germany is illustrated in Figure 4.1.

DS4 workbench illustrated in Figure 4.1 includes four parts. Part 1 is the hydraulic power station, which supplies the hydraulic power to the whole hydraulic system. Part 2 is the hydraulic component arrangement panel. The hydraulic components are arranged on it according to the hydraulic schematic. Part 3 is the electric control panel, which can be used to control the corresponding electrical-hydraulic components according to the electric schematic. Part 4 is the simulation platform which can be used to load the software of Automation Studio. The software can simulate the hydraulic schematic in advance to verify if the circuit design is right.

Software Introduction

Automation Studio (AS) software integrated by simulation platform reproduces the virtual environment of practical equipment operation. Students can design and simulate their circuits through this platform to verify the actual equipment. This platform also can be used to improve the students' thinking innovation ability, which is helpful to better understand the professional knowledge. In the following experiments, we will use this software to verify the experiments beforehand.

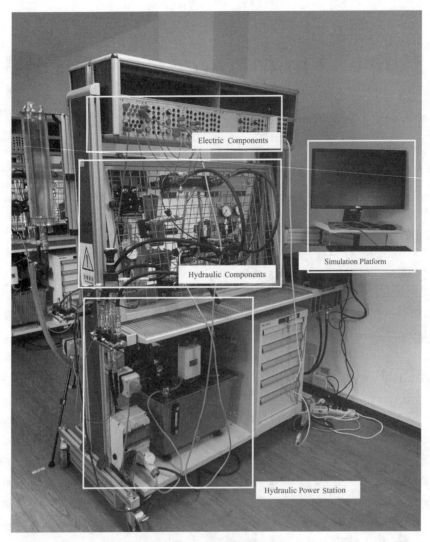

Figure 4.1 Structure of DS4

Automation Studio mechatronics simulation platform is a control-oriented automatic integration system with eight libraries, including hydraulic and proportional hydraulic pressure library, pneumatic pressure and proportional pressure library, electrical control library, digital circuit library, sequence function diagram library, human-computer interaction interface and control panel library, electrician and ladder diagram library. Each component library contains hundreds of symbols which are compatible with ISO, DIN, JIC, IEC and other standards(as Figure 4.2 shows). We can use graphical input in the simulation platform to select the appropriate components and drag them directly to the workspace, click the mouse to connect wires and pipelines automatically, and quickly create any type of automation system. At the same time, the simulation platform can be connected directly with the actual PLC and related electronic control equipment to do semi-physical debugging. Please login the website of https://www.famictech.com/cn/ to get detailed information.

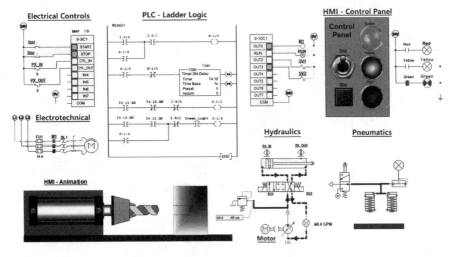

Figure 4.2 Structure of the AS Software

Procedure

This experiment consists of three separate parts. Part 1 is an exercise to select and install a number of fluid power components on the laboratory workstations. Part 2 is symbol identification. Part 3 consists of constructing the components of Part 1 in the software of Automation Studio.

Part 1 Recognize the Components on the Workstation

You will find there are numbered tags on some components of the workstations in the fluid power lab. Choose a workstation from which you will collect information about its components. On each station there are 11 tagged components. On the attached worksheet, fill in your name, lab section, date, and the workstation number that you have chosen.

Component names, manufacturers and model numbers should also be filled in the appropriate locations in Table R.4.1 as Table 4.1 shows.

Table 4.1 Example of Components Information

Tag#	Component Name	Manufacturer	Model Number
1	Directional Control Valve	Rexroth	R900561272

Part 2 Familiar with the Symbol of the Components

Some fluid power symbols are given in Table 4.2, and a list of specifications for several different fluid power components is followed. Select the correct symbol in Table 4.2 and fill its letter designation on the line to the left of its proper description on Table R.4.2 of the lab report. Note that not all of the symbols in Table 4.2 will be used in the matching exercise.

Table 4.2 Fluid Power Symbols

Part 3 Recognize the Circuits

Set up two circuits with different actuators on each stand. Both are common circuits used during the semester. Complete both circuits schematic on the lab report and then

Lab 4 Fluid Power Components and Circuits

reproduce the circuits in Automation Studio and save the files. Your instructor will need to check the circuit and sign off each circuit/student. You will need to run the simulation of each circuit.

Circuit A: This is a cylinder circuit. Draw the circuit and mark the name of the component by the proper tag on the drawing. All the functional components on the workbench should be involved to animate the circuit.

The distribution block on the DS4 workbench has six ports. A manual operated by-pass valve (open/closed) is installed on the block used for start-up of hydraulic power station. When the valve is open, the pump flow goes directly to the tank unrestricted through the by-pass way. After starting up and running the pump, the by-pass valve should be closed for normal operation. The valve tagged 11 is a three-way ball valve.

Circuit B: Replace the cylinder with a hydraulic motor which can be run in bi-directions of rotation.

Note: The directional control valve may be of different spool functions.

Lab 5 Pump Disassembly and Performance

Objectives

(1) Understand how gear, vane, and piston pumps are constructed;
(2) Study how hydraulic fluid is driven through different pumps;
(3) Know common pump specifications;
(4) Understand how pump flow, power and efficiency vary with system pressure;
(5) Learn to build up a hydraulic circuit to testify the related characteristic of the pump;
(6) Use hydraulic components to build up a hydraulic circuit on DS4 experiment bench, and then verify the simulation results from Automation Studio and understand the characteristic of pump.

Equipment Preparation

The experiment makes use of DS4 from Bosch Rexroth of Germany.

Although there are many different kinds of hydraulic pumps, this laboratory exercise will focus on the design and construction of gear, vane and piston pumps. All of these pumps will be available in the laboratory for students to study and disassemble. This is the first part of this lab. The following second part is to study the performance of the vane pump installed on the DS4 workbench.

Part 1 Pump Disassembly

During performing this exercise, students should try to understand the working principle of the pump and know exactly the fluid paths of the pump, how the oil goes through the pump from the low-pressure inlet port to the high-pressure outlet port.

For any questions during this exercise, first read the manufacturer's catalog information provided by the instructor. If you can't find the information in the manufacturer's data sheet, ask the laboratory instructor for help.

There are three different sections in this exercise, each section for one specific

Lab 5 Pump Disassembly and Performance

pump. The order to complete these sections is not critical. For each specific pump, the student should complete a list of specifications and answer a set of questions as outlined in the laboratory procedures.

Students should complete all the three sections in this exercise. Procedures for disassembling the three pumps are given below. For each specific pump, there are corresponding specification tables and question sets in the lab report. These tables and questions should be completed in the laboratory. Copies of manufacturer's data sheets for the pumps used in this exercise will be provided by the instructor. Use this information to assist in determining pump specifications and answering questions.

Procedure

1. Gear Pump Disassembly

(1) Remove the four hex head bolts and open the cap half of the body. Note that the matching punch marks on the end of each casting indicate the proper orientation of the two body halves.

(2) Remove the shaft seal from the mounting face. You can find that all areas inside the pump housing except the high-pressure area are connected to the inlet area by cast-in passages. The leakage oil goes directly to the low-pressure inlet. This internal draining is used in many hydraulic devices to assure that high pressure does not generate in weaker areas of the device (especially around shaft seals), causing damage of the shaft seal or other failures.

(3) Disassemble the built-in relief valve and study how it operates. Some of the gear pumps are not designed with built-in relief valves. If the pump has such a relief valve, you can find it on bottom side of housing after removing the black hex cap.

(4) Reassemble the gear pump in the reverse order of disassembly. Be sure that flange body halves shall be assembled correctly to front half and that O-rings and seals are in place.

2. Vane Pump Disassembly

(1) Remove the four hex head bolts used for holding the pump cover plate in place. Remove the cover plate.

(2) Remove the spring, pressure plate and a-ring, which are located under the covered plate. You can find the oil passages and channels in the pressure plate.

(3) Put the pump in such a direction that the two pins are in the six o'clock and

twelve o'clock positions.

(4) Rotate the pump shaft in the direction indicated by a casting arrow on the outside of the cam ring. Describe the locations of the following in terms of clock position in the data table of the lab report:

①Inlet ports;

②Outlet ports;

③Remove cam ring.

(5)Reassemble the pump in the reverse order of disassembly. Install the cam ring in the opposite direction to the arrow.

3. Piston Pump Disassembly

Compared with the gear pump and vane pump, the piston pump (swash plate axial piston pump as an example) has a more complex structure. So the repair and assembly should be focused on the main pump and variable head.

Pump Body:

(1)Loosen the six hex head bolts that connect the pump body to the variable head, remove the variable head, and place it properly and dust-proof.

(2) Remove the piston and slipper assembly, remove the ball hinge, spring and other components from the drive shaft, and break them down into individual parts.

(3)Remove the cylinder block and its outer insert cylinder sleeve, both for interference with the decomposition.

(4)Remove the valve plate, remove the drive keys, flange screws and flanges and seals, and drive two roller bearing assemblies.

(5) Loosen the connection screw between the pump body and the housing, disassemble the pump body and the housing (but the positioning pin of the flow distribution plate on the pump body can't be removed), and remove the pump body.

(6)Remove the roller bearings.

Variable Assembly:

(1)Remove the variable assembly and remove the thrust plate and pin.

(2)Loosen the lock nut, remove the upper flange, and remove the adjusting screw and the variable piston (for the hydraulic control variable head, disassemble the variable control valve and remove the adjusting screw and variable piston).

(3)Reassemble the piston pump in the reverse order of disassembly. Make sure the order of parts is mutual.

Part 2 Pump Performance

Discussion

Various hydraulic workbench in the fluid power laboratory will be used to demonstrate circuit principles as described in class or laboratory lectures. Each workbench is constructed of several common components. These components include a hydraulic power supply, a flow meter, a manifold on which direction control valves and other components may be mounted, a start-stop switch for the electric motor on the power unit, 24 DC for solenoids or 220 Volt electric devices.

Although each workbench has a set of common components, the components are all unique in terms of their style, size, capacities, and location within the workbench circuit.

All equipment used on the laboratory workbench is industry standard. The equipment that would be used in industrial applications is generally capable of operating at the higher pressure. However, for safety reasons the workbench maximum operating pressure is limited to 50 bar.

Most of the hydraulic power supplies have a relief valve installed on the unit for protection against excessive pressures. This unit has been set and can't be adjusted by the students under any circumstances. If for any reason it becomes necessary to change the setting, the only person authorized to do so is the laboratory technician or the laboratory instructor.

Each of the power unit used in this experiment has been equipped with a valve to restrict the pump output flow. The oil from the pump must flow through this valve before returning to the reservoir. Opening or closing this valve controls the resistance of the oil returning to the reservoir. This resistance results in the development of pressure in the system. In most systems, resistance is caused by loading a cylinder or motor. For obtaining data to plot a pressure versus flow curve for the pump, the restricting valve allows infinite positioning which implies infinite pressure settings. Figure 5.1 is a schematic of a circuit similar to that, which will be built up on Automation Studio. Since this simulation is on the computer, the rotation speed of the pump could be found in software.

When the restricting valve is open, oil will freely flow through the circuit without forming significant pressure. As the valve is closed, the cross-sectional area of the valve orifice is reduced. Thus, the same amount of oil is forced through a reduced area. A higher oil velocity is now needed to force the same oil flow through the partially closed valve. This in turn requires a higher pressure at the pump.

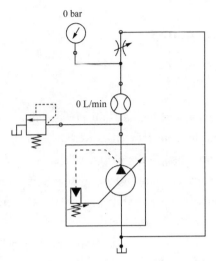

Figure 5.1 Hydraulic Circuit on Automation Studio

To verify the simulation results from Automation Studio, the hydraulic circuit in Figure 5.2 will be built up on DS4 workbench. The measuring glass will be used to measure the volume of the oil within certain time, and then the flow speed of the circuit can be calculated. Additionally, since exact pump shaft speed is not obtainable, it is important to determine theoretical output in another way. The method available to us is to extend our flow versus pressure curve back to the Y-axis (zero pressure). Since the pump theoretically has no internal leakage when its pressure differential is zero, we can assume the theoretical output is the flow with no backpressure. This flow rate is its theoretical output and is what you should use to determine the volumetric efficiency.

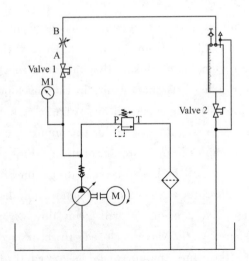

Figure 5.2 Hydraulic Circuit on DS4 Experiment Workbench

Lab 5　Pump Disassembly and Performance

The hydraulic horsepower of the circuit can be calculated using the following equation:

$$\text{Hydraulic Power} = p \cdot q \tag{5-1}$$

where p is the system pressure (bar); q is the actual pump flow (L/min).

The volumetric efficiency of the pump can be calculated using the following equations:

$$\eta_v = \frac{q_a}{q_t} \tag{5-2}$$

where q_a is the actual pump flow; q_t is the theoretical pump flow.

$$q_t = v_d n \tag{5-3}$$

where v_d is the pump displacement (mL/r); n is the pump shaft speed (r/min). (In order to facilitate the engineering data recording, p, q, v_d and n make use of engineering unit in this lab.)

Procedure

1. Building up a Hydraulic Circuit on Software of Automation Studio

In this section, we are going to build a hydraulic circuit to measure the characteristic curve in AS.

Step 1: Drag the elements you need from the library and place them on the diagram as shown in Figure 5.3. The necessary elements are listed in Table 5.1.

Table 5.1　Elements in AS

Element	Ways to Find the Element	Symbol
Pressure Compensated Pump	Hydraulic→Pumps and Amplifiers→Unidirectional Pressure Pump	
Hydrostatic Reservoir(Tank)	Hydraulic→Reservoir→Hydrostatic Reservoir	
Pressure Relief Valve	Hydraulic→Pressure Valves→ Pressure Relief Valves→Relief Valve ISO-1219-1:1991	
Variable Throttle Valve	Hydraulic→Flow Valves→Orifices→Variable Throttle Valve	

(continued)

Element	Ways to Find the Element	Symbol
Pressure Gauge	Hydraulic→Measuring Instruments→Others→Pressure Gauge	0 bar
Flow Meter	Hydraulic→Measuring Instruments→Others→Flow Meter	0 L/min

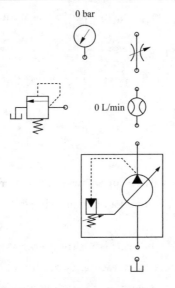

Figure 5.3 Placing the Elements on Diagram

Step 2: Connect the hydraulic elements as shown in Figure 5.1.

Step 3: Double click the pressure relief valve and enter the Data interface. Change the cracking pressure to 50 bar as shown in Figure 5.4.

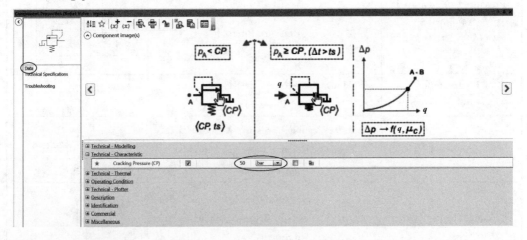

Figure 5.4 Modifying the Cracking Pressure of the Relief Valve

Lab 5 Pump Disassembly and Performance

Step 4: Double click the pump and enter the technical specifications interface. Change the characteristic of the pump so that the specifications of the software coincide with the pump of DS4 workbench (as shown in Figure 5.5).

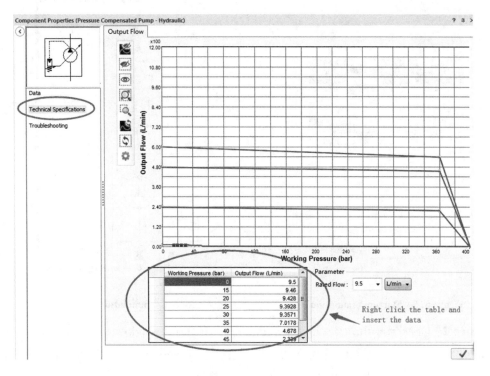

Figure 5.5 Modifying the Characteristic of the Pump

The data of the pump is as shown in Table 5.2.

Table 5.2 Adjusting the Flow of the Pump

Pressure/bar	Flow/(L · min^{-1})	Pressure/bar	Flow/(L · min^{-1})
0	9.5	35	7.017 8
15	9.45	40	4.678
20	9.428	45	2.338
25	9.392 8	50	0
30	9.357 1		

Now, the hydraulic circuit is ready to measure the characteristic curve of the variable pump.

2. Starting Simulation and Measuring the Characteristic Curve of the Pump

In this section, we will first start the simulation of the hydraulic circuit. In the simulation, the flow and pressure will change when you adjust the opening of the throttle valve.

Step 1: Click the simulation on the tool bar and find the Normal Simulation. Clicking the Normal Simulation will start simulation as shown in Figure 5.6 and Figure 5.7.

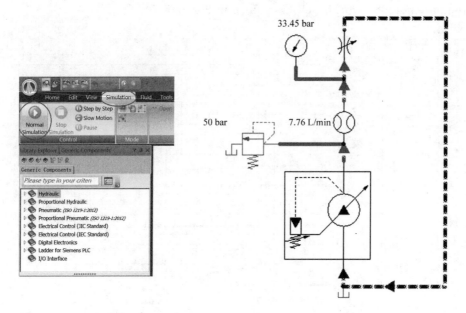

Figure 5.6 Starting Simulation   Figure 5.7 Simulation

Step 2: Click the throttle valve, and there will be a window which allows you to change the opening of the throttle valve as Figure 5.8 shows. Change the opening until the pressure is 15 bar. Record the flow data in lab report Table R.5.1.

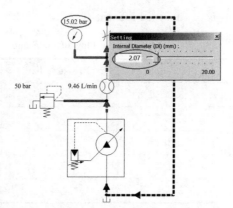

Figure 5.8 Changing the Opening of Throttle Valve

Step 3: Repeat step 2 and change the opening of the throttle valve to change the pressure of the circuit. Record the data in Table R.5.1.

If you want to know more about the variable pump, stop the simulation and double click the pump to enter the data interface as Figure 5.9 shows. You can see the rotating speed of the pump and its displacement. You can also check your results in technical specifications. This part will tell you the characteristic curve of the pump as shown in Figure 5.10.

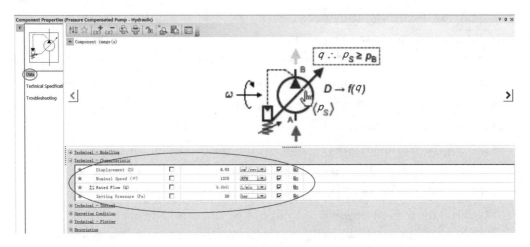

Figure 5.9 Pump Characteristic

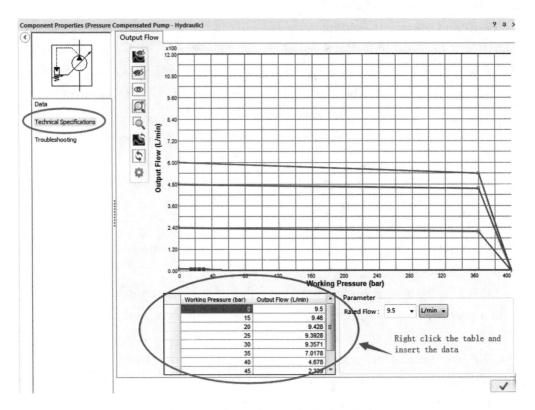

Figure 5.10 Characteristic Curve of the Pump

3. Completing the Experiment on DS4 Workbench and Checking the Simulation Results

A variable displacement pump will be used in this laboratory exercise. A sheet for power unit has been provided for recording data (as shown in Table R. 5. 1). The hydraulic circuit is shown in Figure 5. 2.

(1) In the space provided in the lab report data section, draw a complete schematic of the circuit used for this exercise. Include only those components that are part of this specific test circuit. Physically tracing the circuit will help in drawing it. Record test stand specifications listed on the data sheet.

(2) Note the maximum power output of the electric motor on the test stand. This information will be needed to answer the questions at the end of the lab.

(3) Check the condition of all connections on the test stand.

(4) Check the manual flow control valve to ensure it is fully open.

(5) Jog the pump start switch (quickly turn on and off) to verify that flow occurs without any leaks in the system.

(6) Start the pump and allow it to run continuously.

(7) Record the oil temperature in the reservoir in the appropriate table (as shown in Table R. 5. 1 or Table R. 5. 2). Some of the workbench have thermometers connected into the system lines. Others have a thermometer located on the side of the reservoir.

(8) Close the shut-off valve and adjust the pressure relief valve to a system pressure of 50 bar.

(9) Check the set pressure on the variable pump of the drive power unit (the maximum pressure is 50 bar).

(10) Open shut-off valve. Record pressure and flow data at this moment. Note that the zero pressure will be determined from the plot of the test data as mentioned in the discussion.

(11) Adjust the manual flow control valve until a pressure of 15 bar appears at measuring point.

(12) Close the manual flow control valve slowly to increase the system pressure to the first pressure level attainable noted on the appropriate data table.

(13) Measure the flow through the measuring glass. To do this, close the shut-off valve and close the measuring glass for 15 seconds. Record the pressure and flow. Questions pertaining to reading the flow meters or pressure gages should be addressed to the laboratory instructor.

(14) To facilitate the measurement of the next set of data, open the shut-off valve, release the amount of measuring cup of oil, and then close the shut-off valve.

(15) Increase the pressure to the next pressure level indicated in the data table and

record the pressure and flow.

(16) Repeat procedures 14~15 until flow meter reading drops to zero.

(17) Fully open the manual flow control valve.

(18) Record the reservoir oil temperature.

(19) Shut off the test workbench.

Calculating Power and Efficiency: Use equations (5-1) to (5-3) and the data collected from the test stands to calculate power and efficiency values.

Note: During the calculation, the flow rate should be multiplied by 4 (due to the measured data duration of 15 seconds). Complete Table R. 5. 1. Show all calculation methods in the calculation section of this laboratory instruction.

Plotting Data: Use the data generated from the pumps to plot three performance curves for each pump. These plots must be created in accordance with generally accepted methods. Plots may include more than one curve as long as all information depicted in the plot pertains to one specific pump.

Plot the following information for each of the two pumps used in this exercise. For each plot use pressure as X-axis values:

(1) flow versus pressure;

(2) hydraulic horsepower versus pressure;

(3) volumetric efficiency versus pressure.

Note: When generating plots of the test data, use data from zero pressure through data obtained at the point where flow drops off rapidly. Data beyond this is not an indication of pump performance because either the relief valve is opening or the pump is being destroyed.

Lab 6 Cylinder Circuit Operation

Objectives

(1) Understand common cylinder specifications;
(2) Study normal and regenerative cylinder circuit designs;
(3) Determine experimentally how cylinder circuit design affects cylinder velocity.

Equipment preparation

The experiment makes use of DS4 workbench from Bosch Rexroth of Germany.

Discussion

Hydraulic cylinders are available in many common types and sizes. Cylinders for unique applications can be specially ordered in almost any size. This experiment will be divided into two parts. Part 1 illustrates the normal extending motion cylinder, while part 2 studies the regenerative operation cylinders. Then how cylinder circuit design affects cylinder velocity will be studied.

Figure 6.1 is a sketch of a single end-rod hydraulic cylinder. Several characteristics are used to specify cylinders. However, the three most common parameters used to specify a cylinder are its bore diameter, rod diameter and stroke length. Typically these parameters are typed or stamped on the cylinder in the format of bore diameter×rod diameter×stroke length. For example, $3'' \times 5/8'' \times 12''$ would correspond to a cylinder with a three-inch bore, $5/8''$ diameter rod, and a one foot stroke length.

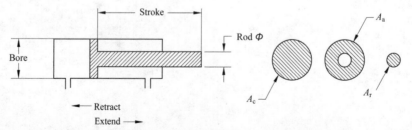

Figure 6.1 A Sketch of a Single End-rod Hydraulic Cylinder

Lab 6 Cylinder Circuit Operation

The velocity of a cylinder is a function of flow and the cylinder area. Three effective areas characterize a cylinder. These are the piston area, annulus area, and rod area. The piston area A_c is defined as the area of the piston of the cylinder. The rod area A_r is only the area of the piston rod. The annulus area A_a is equivalent to the difference between the piston and rod areas.

Part 1 Normal Cylinder Operation

Figure 6.2 (b) is to control the solenoid of Y1, which is used to accomplish normal cylinder operation of Figure 6.2(a) circuit. Under normal operation, cylinder extending and retracting velocities are strictly functions of the pump flow and the cylinder area.

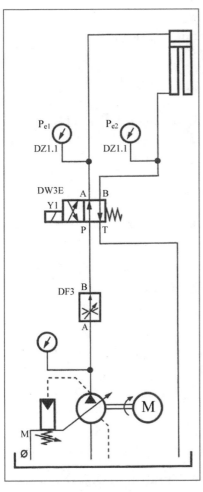

(a) Normal Cylinder Circuit

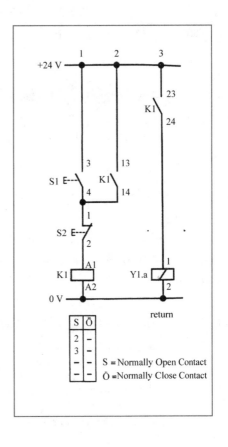

(b) Electrical Control Circuit

Figure 6.2 Schematic of Cylinder Hydraulic System and Its Electrical Circuit

Figure 6.3 shows how the pump flow q_p acts on the piston area A_c, to produce the extending motion of the cylinder. The return flow q_r simply returns to the system's reservoir. In normal retracting motion the flow is simply reversed, and that is, the pump flow enters the rod side of the cylinder and the return flow exits the piston side of the cylinder. The following equations are used to calculate extending and retracting velocities for normal operation.

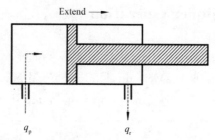

Figure 6.3 Cylinder in Normal Extending Motion

Cylinder extending velocity:

$$v_{ext} = \frac{q_p}{A_c} \tag{6-1}$$

where v_{ext} is the cylinder extending velocity; q_p is the pump flow; A_c is the cylinder piston area.

Cylinder retracting velocity:

$$v_{ret} = \frac{q_p}{A_c - A_r} = \frac{q_p}{A_a} \tag{6-2}$$

where v_{ret} is the cylinder retracting velocity; A_r is the cylinder rod area; A_a is the cylinder annulus area.

The following equations are used to determine the forces that develop during the extending and retracting motions of a cylinder undergoing normal operation.

Extending force:

$$F_{ext} = p \cdot A_c \tag{6-3}$$

where F_{ext} is the cylinder extending force; p is the system pressure.

Retracting force:

$$F_{ret} = p \cdot (A_c - A_r) = p \cdot A_a \tag{6-4}$$

where F_{ret} is the cylinder retracting force.

Procedure-Normal Cylinder Operation

(1) Build up a normal cylinder circuit and electrical circuit as shown in Figure 6.2. Record the experiment workbench number in the report.

(2) Collect all of the information labels on the cylinder and fill them in Table R.6.1 of the lab report.

Lab 6 Cylinder Circuit Operation

(3) Using the information in Table R. 6.1, compute the following:

①Cylinder piston, rod, and annulus areas;

②Expected cylinder extending and retracting velocities based on rated pump flow of 9.5 L/min.

(4) Turn on the pump.

(5) Regulate the opening of flow control valve DF3, and then record the piston side and rod side pressures during cylinder moving and stop, and also record the time from the beginning to the end of extending in Table R. 6.2.

(6) Regulate the opening of DF3 and repeat procedure 5 again, and then record all the information in Table R. 6.2.

(7) Calculate the piston velocity in different openings of DF3 and fill the data in Table R. 6.2.

Part 2 Regenerative Cylinder Operation

Figure 6.4 is a general schematic of a system, which is used to accomplish regenerative cylinder operation. Under regenerative operation, cylinder extending velocity is a function of pump flow, cylinder return flow, and the cylinder area. A cylinder piston velocity is proportional to input flow. With a constant flow to a double acting and single end rod cylinder, piston retracting velocity is greater than piston extending velocity. This is a result of the different volume requirements.

A method of increasing rod extension velocity without increasing pump size will be studied and observed in this exercise. Increasing piston extending velocity is accomplished by using cylinder return flow to supplement pump flow during the extending motion. This process is called regeneration.

Figure 6.5 shows how the return flow q_r is combined with the pump flow q_p, and acts on the piston area A_c, to produce the extending motion of the cylinder. Note that in the regenerative extending motion, none of the working fluid is returned to the reservoir. Also, the entire pump flow effectively acts solely on the rod area of the cylinder. In regenerative operation, piston retracting motion is accomplished in the same manner as in normal operation. That is, the pump flow enters the rod side of the cylinder and the return flow exits from piston side of the cylinder. The following equations are used to calculate extending and retracting velocities for regenerative operation.

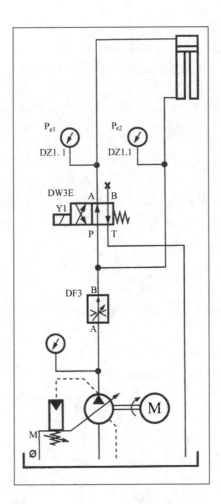

Figure 6.4 Regenerative Cylinder Circuit

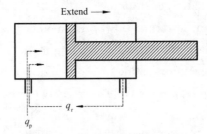

Figure 6.5 Cylinder in Regenerative Extending Motion

Cylinder extending velocity:

$$v_{\text{ext}} = \frac{q_p + q_r}{A_c} = \frac{q_p + v_{\text{ext}} \cdot A_a}{A_c} = \frac{q_p}{A_c} + v_{\text{ext}} \cdot \frac{A_c - A_r}{A_c}$$

Lab 6 Cylinder Circuit Operation

$$v_{ext} = \frac{\frac{q_p}{A_c}}{1 - \frac{A_c - A_r}{A_c}} = \frac{q_p}{A_r} \tag{6-5}$$

where v_{ext} is the cylinder extending velocity; q_p is the pump flow; q_r is the return (exhaust) flow.

Cylinder retracting velocity:

$$v_{ret} = \frac{q_p}{A_c - A_r} = \frac{q_p}{A_a} \tag{6-6}$$

where v_{ret} is the cylinder retracting velocity; A_r is the cylinder rod area; A_a is the cylinder annulus area.

The following equations are used to determine the forces that develop during the extending and retracting motions of a cylinder undergoing regenerative operation. Note that in the extending mode, the pressure applied to the piston area is partially balanced by pressure applied by the fluid on the annulus side of the cylinder (as shown in Figure 6.5). Thus, the extending force is generated by the system pressure acting on an effective area equivalent to that of the cylinder rod area.

Extending force:

$$F_{ext} = p \cdot A_c - p \cdot A_a = p \cdot A_r \tag{6-7}$$

where F_{ext} is the cylinder extending force; p is the system pressure.

Retracting force:

$$F_{ret} = p \cdot (A_c - A_r) = p \cdot A_a \tag{6-8}$$

where F_{ret} is the cylinder retracting force.

Procedure-Regenerative Cylinder Operation

(1) Build up a regenerative cylinder circuit as Figure 6.4 shows. Record the workbench number in the report. The electrical circuit is the same as Figure 6.2(b) shows.

(2) Turn on the pump.

(3) Regulate the opening of flow control valve DF3, and then record the pressure during moving and stop position, and also record the time from the beginning to the end of extending in Table R.6.3.

(4) Regulate the opening of DF3 and repeat procedure 3 again, and then record all information in Table R.6.3.

(5) Calculate the piston velocity in different openings of DF3 and fill the data in Table R.6.3.

(6) Finally, according to the data list in Table R.6.2 and Table R.6.3, draw out the curves in lab report Figure R.6.1 to compare the velocities of different cylinder designs.

Lab 7　Hydraulic Motor

Objectives

Be familiar with the operating features of a hydraulic motor as the following:

(1) Hydraulic motors convert hydraulic energy into mechanical energy (torque and speed);

(2) The direction of rotation or the direction of flow of hydraulic motors can be controlled by means of a directional valve;

(3) The speed of the hydraulic motor is determined by the flow provided and the displacement of the hydraulic motors;

(4) The torque of hydraulic motor is determined by two factors, one is the differential pressure between the inlet and the outlet, and the other is the displacement.

Equipment Preparation

The experiment makes use of DS4 workbench from Bosch Rexroth of Germany.

Discussion

Hydraulic motors are used to produce rotary motion. They can be stalled under load with no damage and are instantly reversible. The rotational speed is directly proportional to the amount of fluid supplied. By changing the input flow, the output speed can be controlled. In this experiment, a standard fixed displacement motor will be operated with variations in the load. Speed will remain nearly constant because the supply is from a fixed displacement pump.

To simulate a load and to show how motor speed may be controlled, flow control valves have been installed in line with the ports of the motor. By increasing the resistance of the motor, we can simulate a load on the motor. This load is developed by the increased pressure required to rotate the motor.

Note: In this exercise, there is no external load applied to the motor shaft. The "virtual" torque you are calculating is actually expended against the motor's own rotor. The energy is consumed in the flow control valve by throttling the "back pressure" on outlet of the motor.

Motor torque:

$$T=\frac{V_d \cdot p}{2\pi} \tag{7-1}$$

where T is the motor torque (N·m); V_d is the displacement of the motor (mL/r); p is the system pressure (Pa).

Flow through motor:

$$q_m=\frac{n \cdot V_d}{231} \tag{7-2}$$

where q_m is the flow through the motor (L/s); n is the motor speed (rmp).

Motor power:

$$P=\frac{T \cdot n}{9\,554} \tag{7-3}$$

where P is the motor power (kW); T is the motor torque (N·m); n is the motor speed (rmp).

Equations (7-1), (7-2) and (7-3) are used to calculate motor torque, flow and power respectfully. (In order to facilitate the engineering data recording, V_d, q_m, n, p make use of engineering units in this lab.)

Procedure

(1) Choose a workbench that is set up with a motor experiment.

(2) A hydraulic schematic circuit is shown in Figure 7.1 and an electric circuit to control the direction valve solenoid is shown in Figure 7.2. Completely assemble the hydraulic and electric circuit on the DS4 workbench, and then check them.

(3) A list of components is given in Table R.7.1 of the lab report for a normal motor test stand. Complete the table with the required data according to the DS4 workbench.

(4) Before you attempt to start the motor on the power supply:

①Record the motor model number in the appropriate locations in Table R.7.1;

②Install a flowmeter and use it to obtain an accurate test result. Make sure the result is displayed in the proper units;

③Check the directional control valve and be certain it is in neutral or central position;

④Check the flow control valves and be certain that they are fully open;

⑤Make sure the by-pass valve is closed. If not, part of the flow would go to tank instead to motor, and the system pressure would be low, even if the motor was stopped.

(5) Start up the power supply motor.

(6) Actuate the directional control valve to operate the motor shaft to rotate clockwise. Record system pressure, flow and motor speed in Table R. 7. 2.

(7) Adjust the flow control valve until the hydraulic motor stops. Record system pressure, flow and motor speed in the zero flow/zero speed row in Table R. 7. 2.

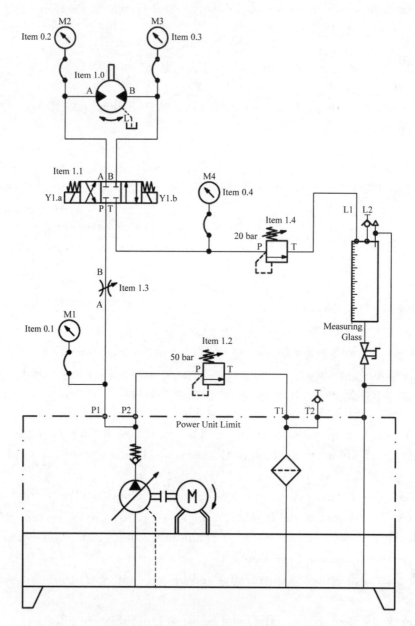

Figure 7.1 Hydraulic Schematic for Motor

(8) In procedures 6 and 7, you have recorded the two operating pressure limitation values for the motor. Calculate four other pressures equally spaced over the pressure range and enter these values in Table R. 7. 2.

(9) Adjust the appropriate flow control valve to obtain the first pressure as determined in step 8. Record the data as before.

(10) Repeat step 9 for the other three pressure values.

(11) Open flow control fully, center the directional control valve and shut off the pump.

(12) Complete Table R. 7. 2 by calculating motor torque, flow, and output power using the equations provided in the discussion of this exercise.

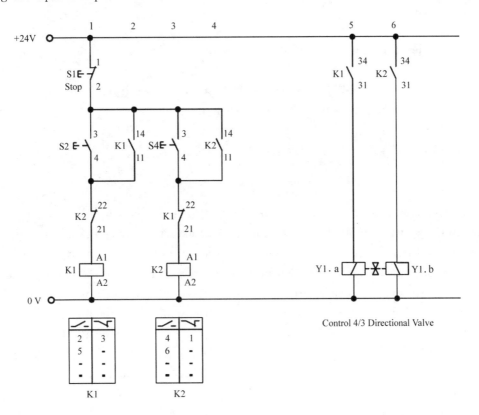

Figure 7.2 Electric Circuit for Motor

(13) Using the data obtained in this exercise, create one plot representing torque versus horsepower for this style of motor tested. Draw smooth curves.

(14) Practice exercise optionally.

Develop circuits with Automation Studio of meter-in, meter-out and a bleed-off type of flow control. Design the circuit as an open loop type of circuit. Animate each circuit and compare with the workbench data. Analysis the differences and reasons.

Lab 8　Valve Disassembly

Objectives

(1) Understand how directional control valves and pressure relief valves are constructed;
(2) Study how hydraulic fluid passes through different valves;
(3) Become familiar with common valve specifications.

Discussion

In this exercise, the design and construction of directional control valves (DCV) and relief valves will be studied. Actual DCVs and relief valves will be available in the laboratory for disassembly. While working on this exercise, focus on understanding the flow paths through the valves. Become familiar with the location of high and low pressure zones in the valve. This will help in understanding how the valves operate.

There are two parts in this exercise: one is the disassembly and assembly of the directional control valve, and the other is the disassembly and assembly of the pressure relief valve. Before disassembling and assembling components, simulate the disassembly and assembly on the mobile phone or computer virtual disassembly platform MLab. After being familiar with the structure and principle of the components, operate on the real components that the MLab assistant provides. The procedures and questions for both parts have been combined in the MLab report section.

Software Introduction

In this experiment, components virtual disassembly and assembly operation will be applied. Log on the following website and download MLab App, and then register according to your name and study ID.

https://www.mools.net/lims/web/down/chemlab.html?from=singlemessage

MLab is a virtual disassembly platform, which is developed by DLUT. There are a lot of subjects except mechanical part. Find out the mechanical part and 11 components are ready to disassembly and assembly (as Figure 8.1 shows). Make sure practice before test.

Lab8　Valve Disassembly

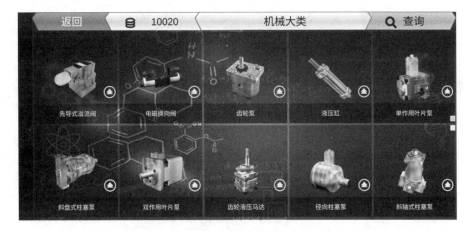

Figure 8.1　Virtual Components

Lab 9 Analysis of Flow Control Valves

Objectives

(1) Learn the valves' flow control function in speed control circuit;

(2) Practice using Automation Studio software to simulate a hydraulic circuit to test the flow characteristic of the flow control valves;

(3) Verify the valves' characteristic simulation result by building up the circuit on the experiment workbench.

Equipment Preparation

The experiment makes use of DS4 workbench from Bosch Rexroth of Germany.

Discussion

The flow in all valves follows basic physical laws.

Valve flow:

$$q = K \cdot \sqrt{\Delta p} \tag{9-1}$$

where q is the flow through the valve; K is a constant for the valve that includes several factors; Δp is the pressure differential between the inlet and outlet of the valve.

Equation (9-1) expresses the relationship of the flow and the pressure in a valve.

When a pressure differential or loss at a particular flow is known, the value of K for the valve can be determined.

Once K is known, it is possible to predict the pressure loss through the valve for any flow. This information may be very important particularly when a valve must be subjected to a flow higher than its rated flow.

This experiment will be divided into two parts. In part 1, a hydraulic circuit simulation model will be built up to measure the flow characteristic of the throttle valve. Additionally, the simulation result will be verified on DS4 experiment bench. In part 2, we will test the speed-regulating valve.

Figure 9.1 is the hydraulic circuit for part 1. Two pressure gauges are set to

Lab 9 Analysis of Flow Control Valves

measure the pressure drop of valve 1. Flow meter measures the flow through valve 1. Valve 2 is used as the loader. By means of adjusting the opening of valve 2, the pressure drop of valve 1 changes, so does the flow. Therefore, the relation between pressure drop and flow can be determined.

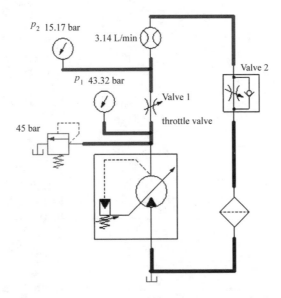

Figure 9.1 Hydraulic Circuit for Part 1 (Throttle Valve)

We just change valve 1 from throttle valve to speed-regulating valve. We can realize the test of speed-regulating valve. Through changing the pressure of hydraulic circuit, the relationship between pressure difference and flow rate of the speed-regulating valve in part 2 can be observed.

Part 1 Throttle Valve

1. Building up a Hydraulic Circuit in the Software of AS

Step 1: Drag the elements you need from the library and place them on the diagram as Figure 9.2 shows. The ways to find the elements are listed in Table 9.1.

Table 9.1　Elements List

Element	Ways to Find the Element	Symbol
Pressure Compensated Pump	Hydraulic→Pumps and Amplifiers→Unidirectional Pressure Pump	
Hydrostatic Reservoir (Tank)	Hydraulic→Reservoir→Hydrostatic Reservoir	
Pressure Relief Valve	Hydraulic→Pressure Valves→Pressure Relief Valves→Relief Valve ISO-1219-1:1991	
Filter	Hydraulic→Fluid Conditioning→Filter	
Variable Throttle Valve	Hydraulic→Flow Valves→Orifices→Variable Throttle Valve	
Variable Non-return Throttle Valve	Hydraulic→Flow Valves→Orifices→Orifice with Non-return Valve→Variable Non-return Throttle Valve	
Pressure Gauge	Hydraulic→Measuring Instruments→Others→Pressure Gauge	0 bar
Flow Meter	Hydraulic→Measuring Instruments→Others→Flow Meter	0 L/min
Variable Flow Controller (This element is only used in part 2)	Hydraulic→Flow Valves→Flow Controls→Variable Flow Controller	

Lab 9 Analysis of Flow Control Valves

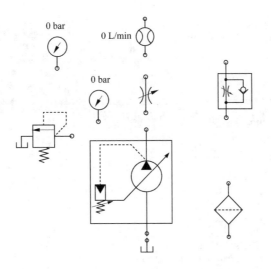

Figure 9.2 Placing the Elements on the Diagram

Step 2: Connect the hydraulic elements as shown in Figure 9.1.

Step 3: Double click the pressure relief valve and enter the data interface. Change the setting pressure to 45 bar as shown in Figure 9.3.

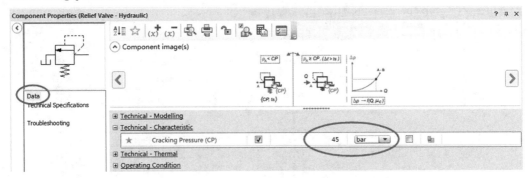

Figure 9.3 Adjust the Crack Pressure of the Relief Valve

Now, the hydraulic circuit of Figure 9.1 has been built up on Automation Studio. The process of simulation will be displayed in procedure 2.

2. Starting Simulation and Measuring the Characteristic Curve of the Throttle Valve

Step 1: Click the simulation on the tool bar and find the Normal Simulation. Click the Normal Simulation and start simulation as Figure 9.4 shows.

Figure 9.4 Starting Simulation

Step 2: Click the variable throttle valve and change its internal diameter to 1 mm as shown in Figure 9.5.

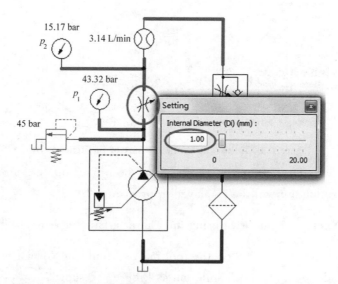

Figure 9.5 Changing the Opening of the Variable Throttle Valve

Step 3: Click the variable non-return throttle valve and change the opening (in Figure 9.6) of it until p_2 is 15 bar, and record the flow rate in lab report Table R.9.1. Just as shown in Figure R.9.5.

Lab 9 Analysis of Flow Control Valves

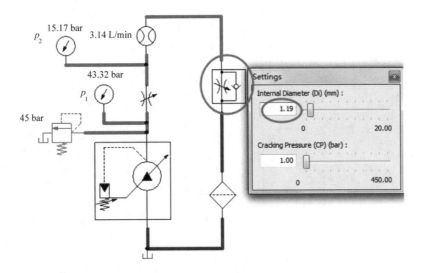

Figure 9.6 Changing the Opening of One-way Throttle Valve

Step 4: Repeat the measurement in step 3 and fill in Table R. 9. 1 in the lab report.

Step 5: Click the variable throttle valve and change its internal diameter to 1.1 mm. Repeat the measurement and fill in Table R. 9. 2 in the report.

3. Throttle Valve Lab Procedure on DS4 Platform

(1) Build up the hydraulic schematic circuit as Figure 9. 7 shows.

(2) Shut off the throttle valve 9 in Figure 9. 7, adjusting relief valve 7 to the pressure $p_1 = 45$ bar.

(3) Turn the throttle valve 9 to scale 3 as Table R. 9. 3 shows.

(4) Make use of throttle valve 8 to adjust the loader pressure $p_2 = 15$ bar.

(5) Read gauge pressure p_1 and p_2 and fill in the blanks of the report form in Table R. 9. 3.

(6) Shut off the measuring glass 3. Shut off valve 2. Record the time until the oil volume goes up to 1 L and fill in the blank of the report form in Table R. 9. 3.

(7) Record the time t when the loader pressure p_2 is 20, 25,30,35,40 bar, and fill in the blank of the lab report form in Table R. 9. 3.

(8) Adjust the opening to scale 4, repeat the procedure above, and then fill the blank in Table R. 9. 4.

(9) Release the relief valve 7, and shut off the electric motor.

(10) Calculate the flow q, and draw the flow characteristic curve in Figure R. 9. 1.

Note: In order to facilitate the engineering data recording, p, q make use of engineering units in this lab.

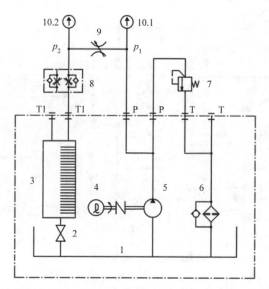

1—Hydraulic Reservoir; 2—Shut-off Valve; 3—Flow Measuring Glass;
4—Electric Motor; 5—Vane Pump; 6—Oil Return Filter; 7—Relief Valve;
8—One-way Throttle Valve; 9—Tested Throttle Valve; 10—Pressure Gauge

Figure 9.7   Hydraulic Circuit on DS4 Bench

Part 2   Speed-regulating Valve

1. Building up a Hydraulic Circuit in the Software of AS

Find variable flow controller from library and replace the throttle valve with speed-regulating valve as Figure 9.8 shows.

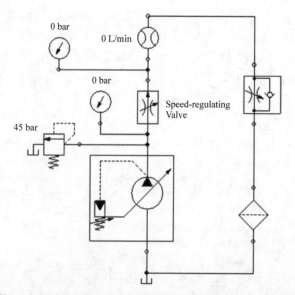

Figure 9.8   Replacing the Throttle Valve by Speed-regulating Valve

2. Starting Simulation of Speed-regulating Valve

Step 1: Click the simulation on the tool bar and find the Normal Simulation as Figure 9.9 shows. Click the Normal Simulation and start simulation.

Figure 9.9 Starting Simulation

Step 2: Click the variable flow controller and change its flow rate to 2 L/min as Figure 9.10 shows.

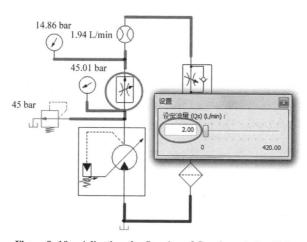

Figure 9.10 Adjusting the Opening of Speed-regulating Valve

Step 3: Click the variable non-return throttle valve and change the opening of it until p_2 is 15 bar, and record the flow rate in Table 9.6 as Figure 9.11 shows.

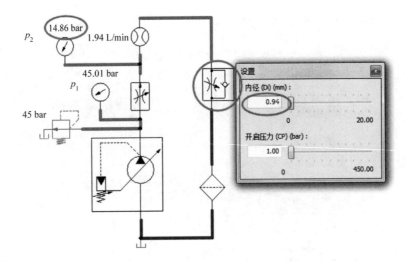

Figure 9.11 Adjusting the One-way Throttle Valve to Change the Pressure

Step 4: Repeat the measurement in step 3 and fill in Table R. 9. 5.

Step 5: Click the variable flow controller and change its flow rate to 6 L/min. Repeat the measurement and fill in Table R. 9. 6.

3. Speed-regulating Valve Lab Procedure on DS4 Platform

(1) According to Figure 9.7, in the place of No. 9, use speed-regulating valve to take the place of throttle valve, and then speed-regulating valve testing stand is built up.

(2) Shut off the throttle valve 8 as Figure 9.11 shows, adjusting relief valve 7 to the pressure $p_1=45$ bar.

(3) Turn the speed-regulating valve 9 to scale 4 as Table R. 9. 7 shows.

(4) Take the use of throttle valve 8 to adjust the outlet of speed-regulating valve loader pressure $p_2=10$ bar.

(5) Read gauge pressure p_1 and p_2 of speed-regulating valve inlet pressure and calculate the pressure differential $\Delta p = p_1 - p_2$, and then fill in Table R. 9. 7.

(6) Shut off ball valve 2 of the flow measuring glass 3, and record the time until the oil volume goes up to 1 L and fill in Table R. 9. 7.

(7) Record the time t when the loader pressure p_2 is 20, 25, 30, 35, 40 bar, and fill in Table R. 9. 7.

(8) Adjust the opening to scale 7 as Table R. 9. 8 shows, repeat the procedures above, and then fill in Table R. 9. 8.

(9) Release the relief valve 7, and shut off the electric motor 4.

(10) Calculate the flow q, and draw the flow characteristic curve in Figure R. 9. 1.

Lab 10   Familiar with Hydraulic Circuits

Objectives

(1) Observe different kinds of hydraulic circuits and become familiar with their functions;

(2) Draw at least four circuits with different functions by using the standard hydraulic component symbols according to the study of the lecture.

Figure 10.1 is the reference figure of hydraulic system. Record the corresponding figures in Figure R.10.1 ~ Figure R.10.4.

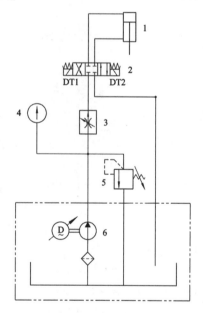

1—Cylinder; 2—Directional Control Valve; 3—Speed-regulating Valve;
4—Pressure Gauge; 5—Pressure Relief Valve; 6—Pump

Figure 10.1   Speed-regulating Circuit

Lab 11　Proportional Control System

Objectives

(1) Understand the working principle of proportional directional valve;
(2) Know the flow characteristic of proportional control valve;
(3) Understand the PID control model and how the proportional gain, integral gain and differential gain influence the close loop control.

Equipment Preparation

The experiment makes use of DS4 workbench from Bosch Rexroth of Germany.

Discussion

The performance of the directional control valves and flow control valves has been presented in the previous experiment. In the previous experiment, you may think that the system is awkward since the valve only has on/off state. In the speed control valve experiment, you may also find out that there is always some deviation when adjusting the opening of the valve to get an ideal system pressure. In this experiment, hydraulic proportional valve will be introduced.

This experiment will be divided into 3 parts. In part 1, the working principle of proportional directional valve will be demonstrated with an open loop hydraulic circuit by AS software. In part 2, there will be a discussion about the flow characteristic of proportional by AS simulation. In part 3, we will briefly explain what is PID control and give you a close loop hydraulic circuit to understand how the coefficients of P, I and D influence the system by AS software. Additionally, a hydraulic circuit will be built up to verify the simulation results on DS4 workbench.

Part 1　Working Principle of Proportional Hydraulic Valve

The cross section of the proportional directional valve is shown in Figure 11.1. The spool is controlled by the electromagnet on the left side. The motion of the spool will

Lab 11 Proportional Control System

influence the opening between P-A and P-B. We will show how it works with Automation Studio next.

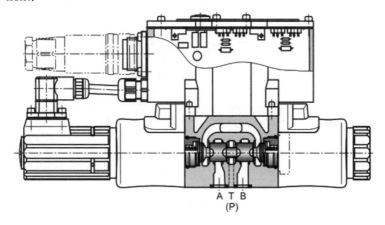

Figure 11.1 Structure of the Proportional Directional Valve

Step 1: Build up a hydraulic circuit. The elements of the circuit are shown in Table 11.1.

Table 11.1 The Elements of the Circuit

Elements	Way to Find the Element	Symbol
Variable Displacement Pump	Hydraulic→Pumps and Amplifiers→Unidirectional Variable Displacement Pump→Variable Displacement Pump	
Hydrostatic Reservoir (Tank)	Hydraulic→Reservoir→Hydrostatic Reservoir	
Pressure Relief Valve	Hydraulic→Pressure Valves→Pressure Relief Valves' Relief Valve ISO-1219-1:1991	
Variable Throttle Valve	Hydraulic→Flow Valves→Orifices→Variable Throttle Valve	
Filter	Hydraulic→Fluid Conditioning→Filter	

(continued)

Elements	Way to Find the Element	Symbol
Proportional Directional Valve	Proportional Hydraulic→Proportional Directional Valves→4 Ports	
Joystick	Proportional Hydraulic→Set Point Device→Joystick	
Double-acting Cylinder	Hydraulic→Actuator→Double-acting Cylinder	

Step 2: Connect the hydraulic elements as shown in Figure 11.2.

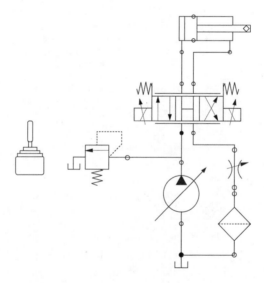

Figure 11.2 Connecting the Hydraulic Elements

Step 3: Double click the proportional directional valve and enter the technical specification interface to change the middle position of the proportional valve to O function as shown in Figure 11.3.

Lab 11 Proportional Control System

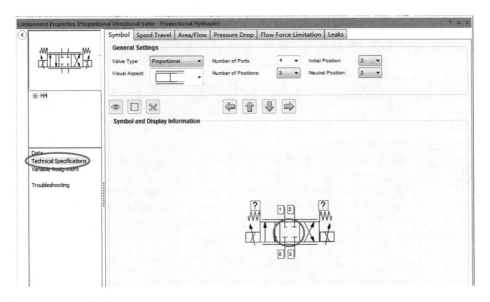

Figure 11.3 Changing the Middle Position of the Proportional Valve

Step 4: Enter the variable assignment interface, and link the joystick with electromagnet as shown in Figure 11.4.

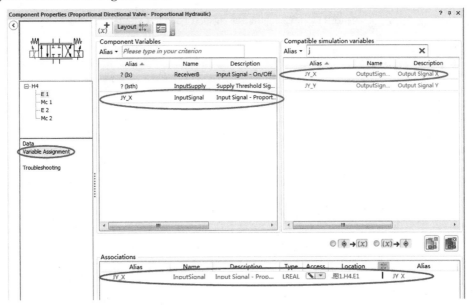

Figure 11.4 Linking the Joystick with Electromagnet

Step 5: The open loop control model has been built up. The stroke of the cylinder is recommended to be changed to 200 mm and the internal diameter of variable throttle valve should be changed to 5 mm (as shown in Figure 11.5 and Figure 11.6). This step can help you observe the principle of the proportional valve.

Note: In order to facilitate the engineering data recording, position, velocity and pressure make use of engineering units in this lab.

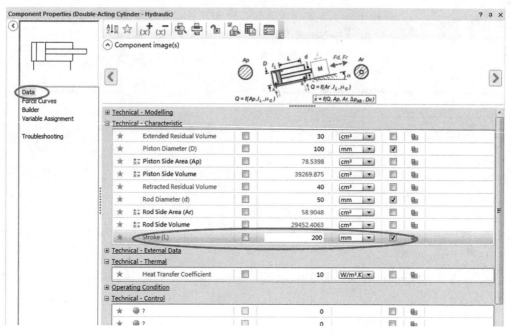

Figure 11.5 Changing the Property of the Cylinder

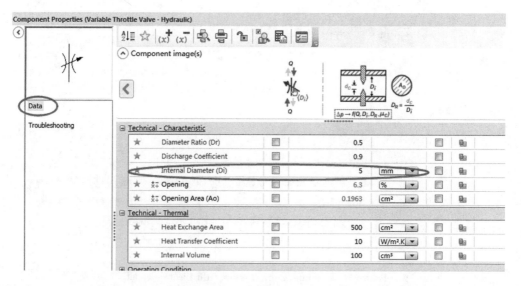

Figure 11.6 Modifying the Property of Variable Throttle Valve

Step 6: Start the simulation. Click the joystick and adjust the X position of the stick. You will find that the spool will move with the joystick as shown in Figure 11.7. Since the spool is only influenced by the joystick and there is no feedback affecting the system, which is a typical open loop system.

Lab 11 Proportional Control System

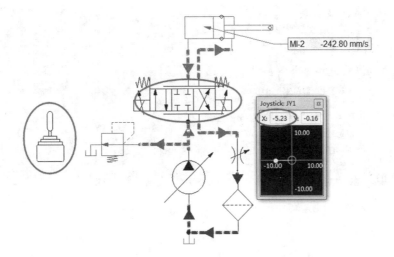

Figure 11.7 Simulation

Part 2 Characteristic Curve of the Proportional Direction Valve

The characteristic curve of the proportional direction valve shows the relation between the flow rate and the spool stroke. The middle position of the valve will have a great influence on the characteristic curve. Figure 11.8 is O type valve and Figure 11.9 is H type valve. In this experiment, we only focus on O function valve.

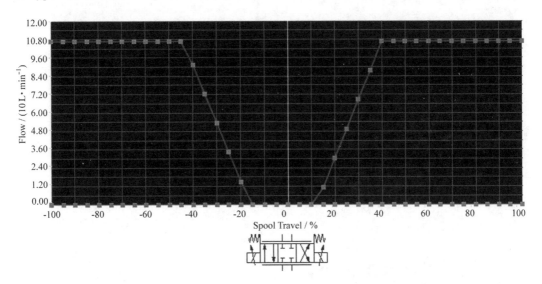

Figure 11.8 Flow Characteristic of Valve with O Function

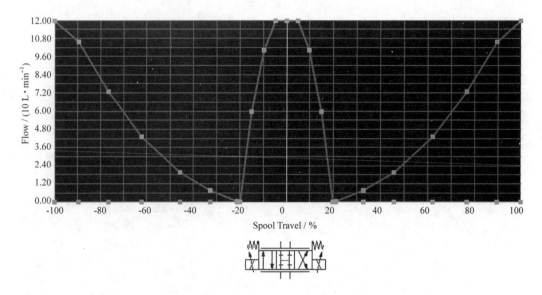

Figure 11.9 Flow Characteristic of Valve with H Function

In general, the characteristic curve of O function proportional directional valve can be summarized by Figure 11.10. For valve with positive overlap, the flow rate stays at 0 although the spool is not at middle position. Zero overlap is an ideal model which has high precision and no leakage. The negative overlap will have some leakages which make the spool not stable. This characteristic can be presented in Automation Studio. s in Figure 11.10 stands for valve spool position, while q_v stands for flow that goes through the valve.

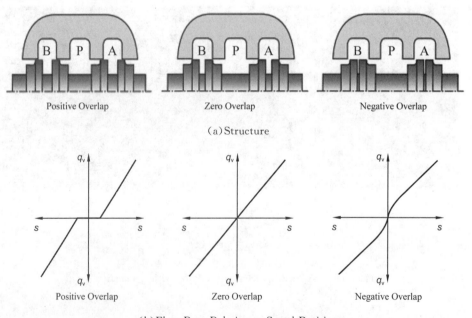

Figure 11.10 Flow Characteristic of Proportional Valve

Simulation

Using the model built up in part 1, change the X position of the joystick to 1.5. You will find the speed of the cylinder remains at 0 mm/s as shown in Figure 11.11. This phenomenon can be illustrated by Figure 11.8 and Figure 11.10. Obviously, the valve in our model is positive overlap!

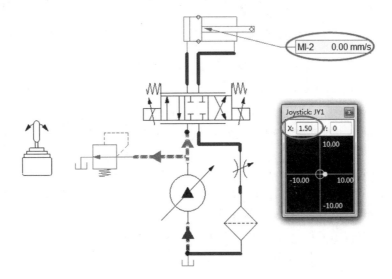

Figure 11.11   The Speed of the Cylinder Remain at 0 mm/s

Part 3   PID Control Simulation and Experiment

PID control is a typical control scheme in the engineering project. In this part, we will briefly talk about how to build a PID close loop control system and try to adjust relevant coefficients of PID to reach our final target. For more theories, you can consult some references about automatic control.

In part 1, you have known that the spool of the directional valve is controlled by the electromagnet. The higher the voltage of the electromagnet, the further the spool travels. Now, try to think about the following question (the voltage limitation is 0 ~ 24 V).

Question: Given a proportional hydraulic valve, how to make the cylinder reach a desired position? In other words, how to make the cylinder extend and then stay at a certain position?

To make the cylinder reach the desired position, the electromagnet should work and make the cylinder extend first. When the cylinder approaches the target position, the electromagnet voltage should be lowered to reduce the speed of the cylinder. Finally, the voltage should be zero when the cylinder reaches the target position.

In summary, the position of the spool is influenced by the error between the desired position and the actual position. We can express the error by the following equation:

$$e(t) = \text{desired position} - \text{actual position} \tag{11-1}$$

where $e(t)$ is the error with respect to time t.

In this case, we can assume that the voltage of the electromagnet can be expressed by the following equation:

$$u(t) = k_1 \times e(t) + k_2 \times \int e(t) \mathrm{d}t + k_3 \times \frac{\mathrm{d}e(t)}{\mathrm{d}t} \tag{11-2}$$

where $u(t)$ is the voltage of the electromagnet; k_1, k_2 and k_3 respectively represent the coefficients of P gain, I gain and D gain.

In equation (11-2), if $k_1 = 1, k_2 = k_3 = 0$, there is only a proportional link. The system is a close loop and only controlled by the proportional link. If $k_1 = k_2 = 1$, $k_3 = 0$, the system is PI system.

These coefficients have different influences on the system response which will be showed to you in the following procedures. For more information about PID control, you can consult some references about automation control.

In the next procedure, we will build a close loop hydraulic system in Automation Studio to solve the problem above.

1. Building a Close Loop Hydraulic System

Step 1: Find the control device from proportional hydraulic → controller → control device(as shown in Figure 11.12).

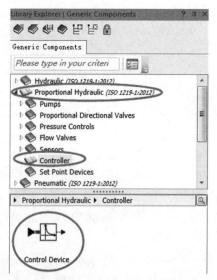

igure 11.12 **Dragging the Controller from Library**

Lab 11   Proportional Control System

Step 2: Connect the elements as shown in Figure 11.13.

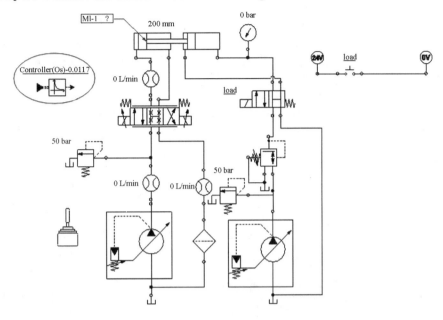

Figure 11.13   Connecting the Elements

Step 3: Double click the control device, enter the output variable interface and change the output to JY X-position (as shown in Figure 11.14).

Figure 11.14   Setting the Output Signal on Controller

Step 4: Link the control device with the proportion direction control valve like part 1(as shown in Figure 11.15). Now a close loop hydraulic system is built up.

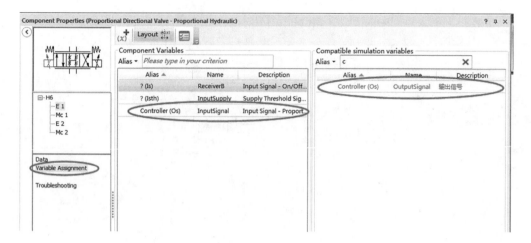

Figure 11.15 Linking the Controller with the Proportional Valve

2. Changing the PID Coefficient

Step 1: Start the simulation. Left click the controller and change the coefficient of the proportional gain from 1 to 5(as shown in Figure 11.16).

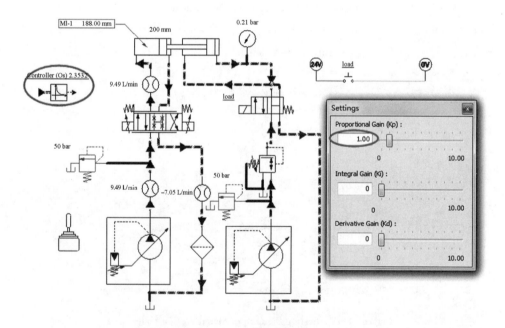

Figure 11.16 Adjusting the Coefficient of Controller

Step 2: Changing the proportional gain to 8, oscillations will happen and the system is not stable. You can use the plotter tool to help you observe the response curve as shown in Figure 11.17.

Lab 11 Proportional Control System

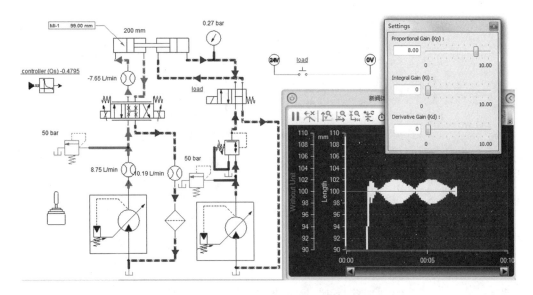

Figure 11.17 Changing the Proportional Gain to 8

Step 3: Change the coefficient to $k_1=1$, $k_2=0.5$, $k_3=0.3$, and overshoot happens. You can observe the response curve as shown in Figure 11.18.

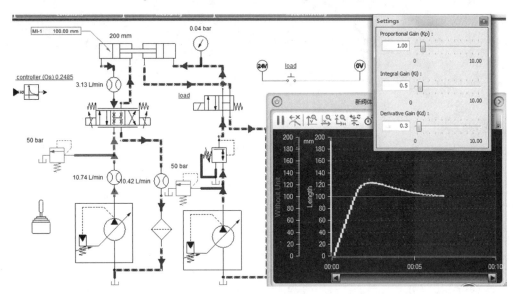

Figure 11.18 Observing the Response Curve

You can try other coefficient to feel the PID control scheme.

Step 4: Press the load button to apply the load on the cylinder as shown in Figure 11.19. Due to the close loop control of the proportional valve, the cylinder almost stays at the same position. As K_p increases, the error will decrease.

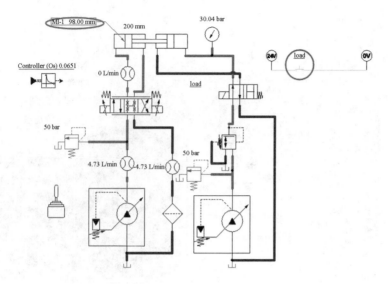

Figure 11.19   Applying Load on the Cylinder

3. Building a Close Loop System on DS4 Workbench

The hydraulic circuit on DS4 workbench is as shown in Figure 11.20. The proportional valve can control the extension and retraction of the cylinder. Through controlling the voltage of electromagnet Y1.b, the cylinder can reach any position which is set in advance.

The right part of the circuit is a load system. Load will be applied on cylinder Pos 1.0 through the cylinder Pos 2.0. The size of load can be adjusted by the reducing valve.

Figure 11.21 is the electrical circuit diagram of this close loop. Figure 11.22 is the electrical circuit of the system. Switch S1 is the main switch controlling the power of the electrical circuit. Switch S2 is the load switch. Switch S3 can shift the controller from setting mode to close loop mode. Switch S4 can shift the controller from close loop mode back to setting mode. Switch S5 can extend the cylinder to position 1(10 mm). Similarly, switch S6 can extend the cylinder to position 2(140 mm).

Steps to conduct the experiments on DS4 platform:

(1) Label the calibration of the position sensor according to Table 11.2.

(2) Turn on the pump. Close the by-pass valve and adjust the setting pressure of relief valve(Pos.1.2) to 50 bar.

(3) Adjust the setting pressure of reducing valve (Pos.2.2) to 30 bar.

(4) Turn switch S1 on. The electrical circuit will be powered on.

(5) Turn switch S4 on to switch the controller to setting mode. Adjust the proportional gain K_p to 1.

(6) Turn switch S3 on to switch the controller back to close loop mode.

(7) Turn switch S5 on to extend cylinder to 10 mm.

Lab 11 Proportional Control System

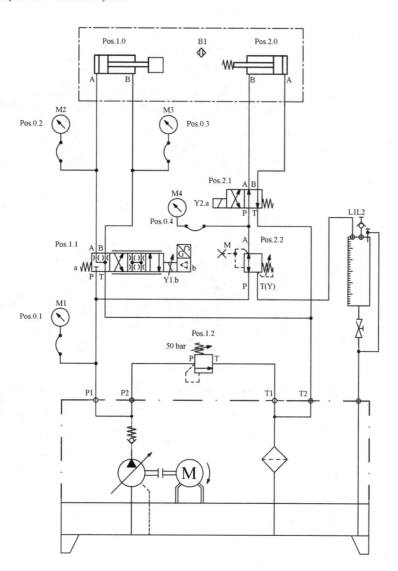

Figure 11.20 Hydraulic Circuit on DS4 Workbench

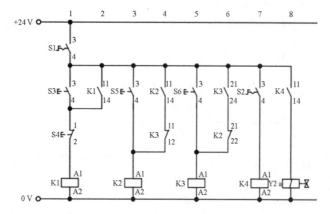

Figure 11.21 Electrical Circuit Diagram

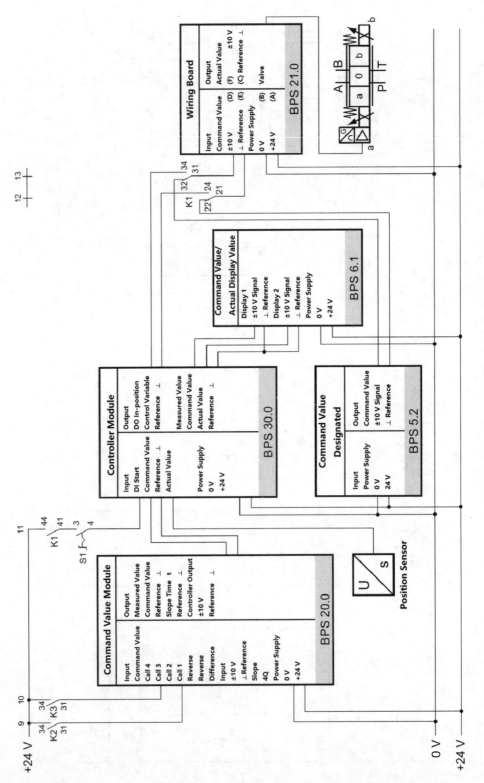

Figure 11.22 Electrical Circuit of the System

Lab 11 Proportional Control System

(8) Turn switch S6 on to extend cylinder to 140 mm. Record the time it takes and whether the cylinder oscillates or not and record the data in lab report Table R.11.1.

(9) Turn switch S2 on to apply the load on cylinder. Record the position of the cylinder and the actual voltage value of the position sensor in lab report Table R.11.2.

(10) Turn switch S5 on to retract the cylinder. Turn S4 on and adjust the proportional gain K_p to 2.

(11) Repeat the steps 1~8 and compare the results in Table R.11.1 and Table R.11.2 with Table 11.3 and Table 11.4. Analyze the error and give the reasons.

Table 11.2 Label the Calibration of the Position Sensor

Position/mm	Voltage/V	Position/mm	Voltage/V
1	0.03	75	2.25
5	0.15	80	2.4
10	0.3	85	2.55
15	0.45	90	2.7
20	0.6	95	2.85
25	0.75	100	3
30	0.9	105	3.15
35	1.05	110	3.3
40	1.2	115	3.45
45	1.35	120	3.6
50	1.5	125	3.75
55	1.65	130	3.9
60	1.8	135	4.05
65	1.95	140	4.2
70	2.1	—	—

Table 11.3 Reference Experiment Results without Load

Proportional Gain K_p	Position/mm	Actual Valve/V	Error/V	Note
1	138	4.14	0.06	Extend Slow
2	138	4.14	0.06	Extend Slow
5	140	4.19	0.01	Extend Fast
6	140	4.20	0	Fast but Overshoot
8	140	4.20	0	Oscillation Unstable
10	140	4.20	0	stable if $K_t=1$ Unstable
12	140	4.20	0	stable if $K_t=1$

Table 11.4 Reference Experiment Results under Load

Load/bar	Position/mm	Actual Value/V	Load/bar	Position/mm	Actual Value/V
0	140	4.20	25	140	4.20
5	140	4.20	30	140	4.19
10	140	4.20	35	139.5	4.19
15	140	4.20	40	139	4.19
20	140	4.20			

Lab 12　Hydraulic Throttle Control Experiment for Engineering Machinery

Objectives

(1) Recognize the hydraulic components in engineering machinery, and learn to correctly connect the hydraulic by-pass throttle control circuit;

(2) Understand the working principle of hydraulic throttle control circuit of engineering machinery;

(3) Understand the performance of the actuator in single or parallel operation.

Equipment Preparation

This experiment makes use of WS290 workbench from Bosch Rexroth of Germany.

Figure 12.1 is its structure. The differences between WS290 and DS4 are the hydraulic power stations and their hydraulic components. WS290 uses the hydraulic pump with load-sensing control (LS) instead of pressure-control, which can be used for experiments in typical engineering machinery application. On the other hand, the hydraulic components for them are also different, for example, WS290 includes some typical control blocks and corresponding valves for engineering machinery.

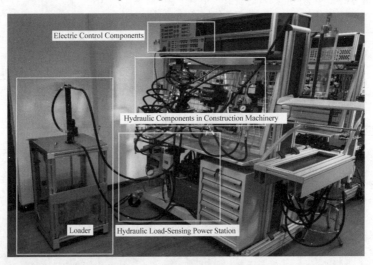

Figure 12.1　Structure of WS290

Principle

1. Working Principle of By-pass Throttle Speed Control Circuit

(1) Throttle Speed Control Circuit with a Single Actuator

As shown in Figure 12.2, there are two throttle orifices on both circuits to actuator and by-pass to tank. The openings of these orifices are linked to each other because they are on the same spool. When we increase the opening of orifice in the oil circuit to actuator (called "working orifice"), the opening of orifice in the oil circuit of by-pass (called "by-pass orifice") will be decreased and vice versa. If one of the orifices is completely opened, the other one will be completely closed. When the inlet pressure of the by-pass orifice is greater than the load pressure of the actuator, part of the oil from the hydraulic variable displacement pump will open the load-holding valve and go to the actuator through the working orifice, and the remaining oil of the pump will go back to the tank through the by-pass orifice at the same time. So the oil flow to the hydraulic cylinder (or the speed of the cylinder) can be controlled by adjusting the opening area of the working orifice in the circuit to actuator. This kind of hydraulic circuit is called the by-pass throttle speed control circuit. In the circuit, the pressure relief valve in Figure 12.2 is used as a safety valve and is normally-closed (only opened when the system pressure is overloaded, so its setting pressure is generally 1.1~1.2 times of the maximum working pressure). The output pressure of the hydraulic variable displacement pump varies with the external load of the actuator.

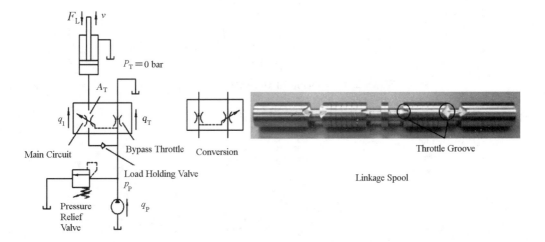

Figure 12.2 By-pass Throttle Speed Control Circuit of a Single Actuator

q_p is the output flow of hydraulic pump (L/min); p_p is the output pressure of hydraulic pump (bar); q_1 is the oil flow to the hydraulic cylinder (L/min); A_T is the opening area of the working orifice (mm^2); q_T is the flow through the by-pass orifice (L/min); v is the speed of the hydraulic cylinder (m/s); F_L is the force of external load (N).

The control method (as shown in Figure 12.2) is also called open center control, and the working principle is shown in Figure 12.3. With the movement of spool, the opening of working orifice to the actuator will be increased and finally the orifice will be fully open. The process can be explained by four stages.

Stage (1): The working orifice to the actuator is fully closed, and by-pass orifice is fully open. At this time, all the oil output from the hydraulic pump goes through the by-pass to the tank. This stage is called the "by-pass back to the tank" and corresponds to the spool stroke 0% phase as shown in Figure 12.4.

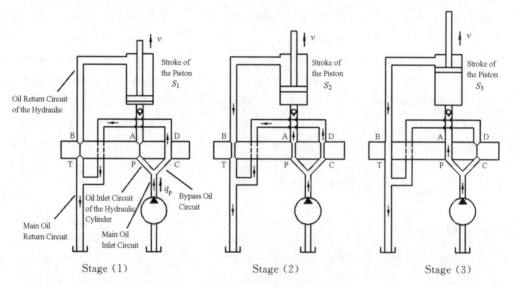

Figure 12.3   Principle of Open Center Control System

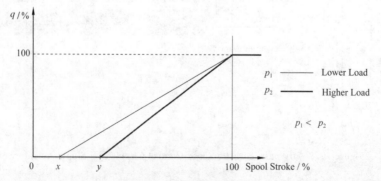

Figure 12.4   Relationship between Spool Stroke of Directional Valve and Flow-rate under Different Loads

Stage (2): The working orifice to the actuator side is gradually open, and the by-pass orifice is gradually closed. The directional valve spool changes from the neutral position to a certain stroke (at this stroke, the hydraulic cylinder just starts to move). That is to say, the hydraulic cylinder does not move with the increase of the directional valve spool stroke, which is called the "dead zone" and corresponds to the spool stroke $0-x\%$ (lower load) or $0-y\%$ (higher load) phase as shown in Figure 12.4.

In stage (2), if the pressure in return line is 0 bar, the output pressure of the hydraulic pump is equal to the pressure differential across the by-pass orifice. According to the thin-walled orifice flow equation (12-1), the pressure differential across the by-pass orifice Δp is shown in equation (12-2):

$$q = CA\sqrt{\frac{2\Delta p}{\rho}} \tag{12-1}$$

$$\Delta p = \frac{\rho q^2}{2C^2 A^2} \tag{12-2}$$

where q is flow through the by-pass orifice (L/min); C is discharge coefficient (C is 0.62 in this experiment); $A = \frac{\pi d^2}{4}$ is section area of orifice (mm^2); Δp is pressure differential of the by-pass orifice (bar); ρ is oil density (kg/m^3).

In order to facilitate the engineering data recording, q, A, P make use of engineering units in this lab.

With the gradual reduction of throttle port area of A in the oil circuit of by-pass, the pressure differential Δp of the by-pass orifice increases gradually, until Δp equals to the load on the hydraulic cylinder. At this time, the oil starts to push the cylinder piston up, and the oil in working circuit to actuator starts to flow. See stage (3).

Stage (3): At this time, part of the oil flows to the hydraulic cylinder, the piston of the hydraulic cylinder goes up, and the other part of oil flows back to the oil tank through the by-pass orifice. Thus in this stage, with the further increase of the spool stroke, the oil flow-rate to the hydraulic cylinder will gradually increase. So this stage is called "precision control area" and corresponds to the x-100% (lower load) or y-100% (higher load) phase of spool stroke as shown in Figure 12.4.

Stage (4): The working orifice to the actuator is fully open, the by-pass orifice is fully closed at the same time, and all the oil flow goes to the hydraulic cylinder. This stage is called "full load" and corresponds to the 100% phase of spool stroke as shown in Figure 12.4.

S_i is the stroke length of the cylinder piston, $S_1 < S_2 < S_3$.

(2) By-pass Throttle Speed Control Circuit with Double Actuators

In the actual operation, there are normally two or more actuators working simultaneously in the circuit, as shown in Figure 12.5, which stands for the by-pass throttle speed control circuit with two actuators. In this circuit, the two actuators are operated in parallel. When two actuators under different loads are working at the same time, the oil flow will go to the actuator under lower load preferentially. That is, at the same spool stroke of the directional valve corresponding to the two actuators, the actuator under lower load will move faster than the actuator under higher load. When different actuators are moving simultaneously, the initial spool stroke is also different when the actuators start running under different loads. That is, the "starting point" of the actuator will be dependent on the load and present a positive correlation. Figure 12.4 shows the relationship between the flow obtained by two actuators under different loads and the spool stroke of the directional valve in the working process. As it can be concluded from the figure, the spool stroke for starting point of the actuator that subjects to the lower load 1 is smaller than the spool stroke for starting point of the actuator that subjects to the higher load 2.

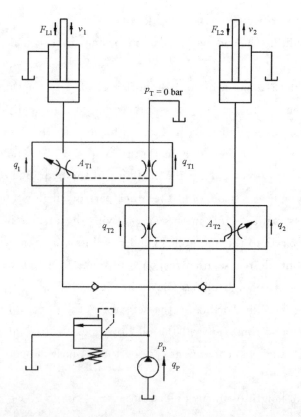

Figure 12.5 Schematic Diagram of By-pass Throttle Speed Control Circuit of Double Actuators

Lab 12 Hydraulic Throttle Control Experiment for Engineering Machinery   81

(3) Cavitation Phenomenon of the Hydraulic Cylinder

Because of the vertical installation of the hydraulic cylinder, the 80 kg load will make the down movement of hydraulic cylinder rapidly, thus forming a negative pressure in the piston side of the hydraulic cylinder. This phenomenon is called cavitation phenomenon. This type of load is called "negative load". All the output oil of the hydraulic pump enters the hydraulic cylinder in this case.

During the experiment, the manual operation of directional valve spool will move the hydraulic cylinder to go down, and the hydraulic cylinder will drop rapidly after the "starting point", at which time the pump reaches the maximum output flow. Then the loader will contact and press the spring. As the spring compresses, the loader will have a drop limiting point, and the cavitation process ends. The spring bounces the loader up, and the loader and spring return to equilibrium quickly. After a brief pause in the equilibrium position, the oil will reestablish pressure in the piston side of the cylinder. When the pressure can push the loader down further, it will continue to go down until the spring is compressed to its limitation, and the drop process is ended. (Cavitation phenomenon can be eliminated by installing a counterbalance valve in the oil return circuit during piston fall.)

2. Structure and Working Principle of Control Block SM-12

In this experiment, the control block SM-12 of Bosch Rexroth is used. The external structural set-up of SM-12 is shown in Figure 12.6. The cross-section and schematic circuit of directional valve in SM-12 are shown in Figures 12.7 and 12.8, and the technical data of SM-12 is shown in Table 12.1.

Control block SM-12 is mainly composed of an oil inlet valve section, some directional valve sections and an oil outlet valve section.

The directional valve is designed according to the 6-way principle, as shown in Figure 12.8.

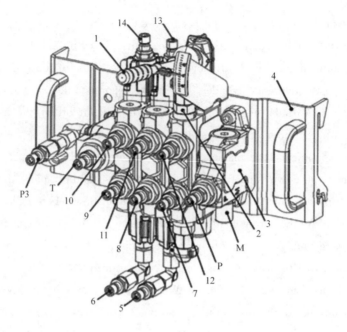

1—Control Lever with Display; 2—Sealed Secondary Pressure Relief Valve; 3—Input Element; 4—Retaining Sheet; 5—Pilot Oil Port "b" of Directional Valve Disc 2; 6—Pilot Oil Port "b" of Directional Valve Disc 3; 7—Working Port "B" of Directional Valve Disc 1; 8—Working Port "B" of Directional Valve Disc 2; 9—Working Port "B" of Directional Valve Disc 3; 10—Working Port "A" of Directional Valve Disc 3; 11—Working Port "A" of Directional Valve Disc 2; 12—Working Port "A" of Directional Valve Disc 1; 13—Pilot Oil Port "a" of Directional Valve Disc 2; 14—Pilot Oil Port "a" of Directional Valve Disc 3; M—Manometer Connection; P—Port for Pressure Line; T—Port for Relief Line; P3—Pressure Port

Figure 12.6 External Structural Set-up of SM-12

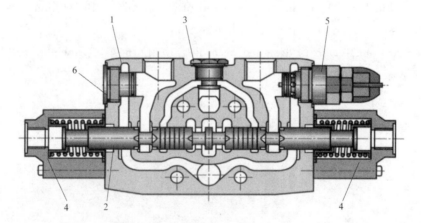

1—Valve Body; 2—Spool; 3—Check Valve; 4—Executive Valve; 5—Second Valve; 6—Plug Screw

Figure 12.7 Cross-section of Directional Valve in SM-12

Lab 12 Hydraulic Throttle Control Experiment for Engineering Machinery

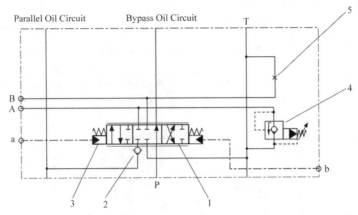

1—Spool; 2—Check Valve; 3—Excutive Valve; 4—Second Valve; 5—Plug Screw

Figure 12.8 Schematic Circuit of Directional Valves in Parallel

Table 12.1 Technical Data of Control Block SM-12

Denomination	Unit	Value
Max. Pressure	bar	120
Max. Control Pressure	bar	30
Input Element:		
LS Pressure Limitation	bar	100
1st Directional Valve:		
LS Pressure Limitation	bar	100
Secondary Pressure Limitation	bar	50
Max. Flow from P-A; from P-B	L/min	70
Reduction via Stroke Limitation		Maximum Value upon Delivery
Type of Actuation		Hydraulic
2nd Directional Valve/3rd Directional Valve:		
Secondary Pressure Limitation	bar	Without
Max. Flow from P-A; from P-B	L/min	70
Type of Actuation:		Hydraulic
Operating Temperature		20~60 ℃
Hydraulic Connections	Quick Release Coupling with Connector, Type W	
Manometer Connections	Threaded Coupling AB20-11/K1 G 1/4	

Preparation

The workbench adopts the workbench of Bosch Rexroth WS290. The schematic hydraulic system circuit is shown in Figure 12. 9. And the diagram of components connection in the WS290 workbench is shown in Figure 12. 10. The list of hydraulic components required for the experiment is shown in Table 12. 2.

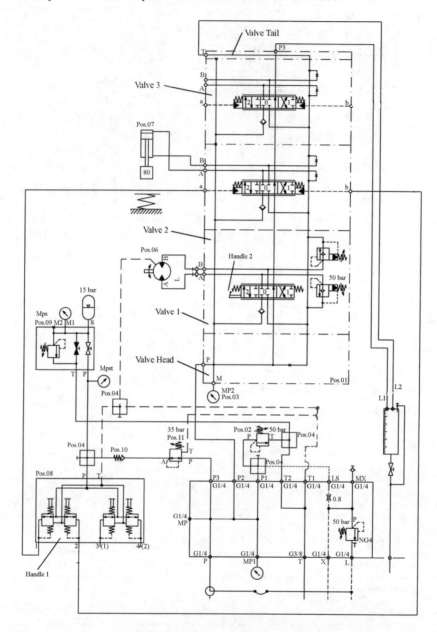

Figure 12. 9 Schematic Circuit of Hydraulic Throttle Control System

Lab 12 Hydraulic Throttle Control Experiment for Engineering Machinery

Table 12.2 **List of Components**

Position	Number	Component Name	Model
01	1	Traveling Machinery Control Block	3SM-12
02	1	Pressure Relief Valve	DD1.1N
03	2	Pressure Gauge with Hose and Quick Release Coupling without Check Valve	DZ1.4
04	4	Distributor Plate with Four Ports	DZ4
05	2	Flow-meter	DZ30
06	1	Hydraulic Motor	DM8
07	1	Double-acting Cylinder with Single-sided Piston Rod with Load	ZY
08	1	Handle	2-2TH6
09	1	Accumulator Safety Block for Diaphragm-type Accumulator	DZ3.2
10	1	Check Valve	DS2.1
11	1	Reducing Valve	DD2.N
12	Several	Hose with Quick Release Coupling with Check Valve	DZ25.1
13	Several	Hose	VSK 1

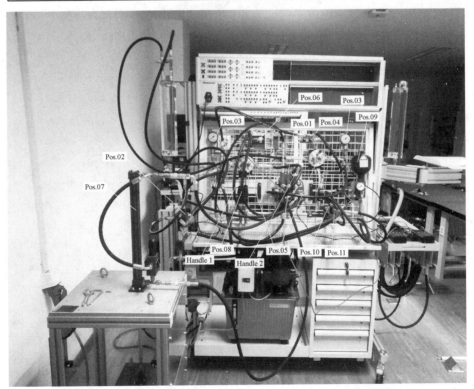

Figure 12.10 **Diagram of Components Connection in the WS290 Workbench**

Procedure

(1) Connect the circuit correctly on the workbench WS290 according to Figure 12.9 and Figure 12.10, and then check it.

Note the operating requirements of LS pump station(LS: Load Sensing).

Oil inlet p and oil port LS must be connected. If there is no pressure at oil port LS, the pump outlet pressure p is the same as the spring pressure of displacement regulator, about 15 bar, which cannot provide the pressure required for the normal use of the system.

(2) Open the workbench. Insert the power plug, turn on the switch button of workbench master, turn on the switch button of hydraulic pump, and rotate the 3-way ball valve to the working position (handle level). Then the hydraulic oil is normally supplied, and the workbench is ready to operate.

(3) Supply the control oil. The accumulator is charged by closing the manual relief (hand wheel) at Pos. 09, and the pressure relief valve DD2 at Pos. 11 is set at 35 bar.

(4) Operate singly. In order to calibrate the moving direction of the actuator, handles 1 and 2 are operated separately to make the hydraulic cylinder and hydraulic motor run in different directions. Observe the up and down of the hydraulic cylinder and the clockwise and anti-clockwise rotation of the hydraulic motor. When the hydraulic motor rotates clockwise, record the return oil of by-pass circuit and T circuit at different rotating speeds (record five groups), observe the change of oil flow through the by-pass circuit and T circuit in the measuring gauge, and then record the results in Table R. 12.1 of the lab report and answer the corresponding questions.

(5) Operate parallelly 1. First of all, let the hydraulic motor rotate at 50% of maximum speed by operating handle 2 and keep the deflection angle of handle 2 unchanged (The handle position is shown in Figure 12.10). Then let the hydraulic cylinder rise by operating handle 1. Observe the change of the speed of hydraulic motor in this process, and answer the corresponding questions in the report.

(6) Operate parallelly 2. First of all, operate handle 1 to raise the piston of the hydraulic cylinder to the maximum height, and then operate handle 2 to set the hydraulic motor at the maximum speed, and finally operate handle 1 to lower the piston of the hydraulic cylinder. Pay attention to the effect of cavitation phenomenon on the falling process of the piston, and observe the variation of the falling speed of the piston and speed variation of the hydraulic motor in this process. Answer the corresponding questions in the lab report.

(7) Operate parallelly 3. First of all, let the hydraulic motor rotate at 50% of maximum speed by operating handle 2, and then operate handle 1 to make the hydraulic

cylinder slowly rise or fall. In this process, try to operate handle 2 to keep the speed of the hydraulic motor unchanged, and answer the corresponding questions in the lab report.

(8) Close the workbench. Put the actuator in the no-load state, rotate the 3-way ball valve to the closed position (handle vertical), control the directional valve spool for several times to remove the remaining pressure in the control block, close the hydraulic pump, open the manual relief (hand wheel) to discharge the accumulator, and turn off the master switch button of the workbench and pull out the power plug.

(9) Clean the workbench, and then the experiment is finished.

Lab 13 Hydraulic LS Control Experiment for Engineering Machinery

Objectives

(1) Recognize the hydraulic components in engineering machinery, and learn to correctly connect the hydraulic load-sensing(LS) control circuit;

(2) Understand the working principle of hydraulic LS control circuit of engineering machinery;

(3) Understand the performance of the actuators in the LS control system under the conditions of sufficient flow and insufficient flow.

Equipment Preparation

This experiment makes use of WS290 workbench from Bosch Rexroth of Germany. Detail information is shown in Figure 12.1 of lab 12.

Principle

1. Working Principle of LS Control Circuit

It can be seen from the thin-walled orifice flow equation (13-1) that when the pressure differential Δp is constant, the flow q through the throttle valve is constant. The LS control circuit is shown in Figure 13.1. The circuit transmits the load pressure signal to the displacement regulator of the hydraulic variable displacement pump to make the pressure differential across the throttle valve constant, thereby eliminating the effect of load changes on the speed of the actuator. This pressure signal from the load to the pump regulator is called a load-sensing signal, also known as the LS signal. This hydraulic control circuit is called an LS control circuit. The use of a hydraulic variable displacement pump can reduce the power consumption of the hydraulic system.

$$q = CA\sqrt{\frac{2\Delta p}{\rho}} \tag{13-1}$$

where q is flow rate in the orifice (L/min); C is discharge coefficient (C is 0.62 in this experiment); $A = \dfrac{\pi d^2}{4}$ is section area of orifice (m²); d is throttle valve opening (mm); Δp is pressure differential at both ends of orifice (bar); ρ is oil density (kg/m³).

In order to facilitate the engineering data recording, q, p, A, d make use of engineering units in this lab.

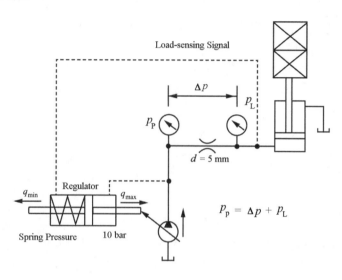

Figure 13.1　LS Circuit of a Single Actuator

(1) LS Control Circuit for a Single Actuator

In the LS control circuit of a single actuator, the load pressure is transmitted to the pump regulator through the oil signal line, which in turn regulates the output flow of the hydraulic variable displacement pump according to the change of the load pressure. As shown in Figure 13.1, the spring side of the regulator is connected to the load oil line, and the right side is connected to the output pressure of the hydraulic variable displacement pump. The load pressure and the spring pressure of the regulator work together on the side of the adjustment plunger, and the output pressure of the pump works on the other side of the regulating plunger with the same working area. In steady operation, the sum of the load pressure and the spring pressure of regulator are equal to the output pressure of hydraulic variable displacement pump. This satisfies the equation (13-2):

$$p_L \cdot A_c + F = p_p \cdot A_c \tag{13-2}$$

where F is actual spring force (N); p_p is output pressure of hydraulic pump (bar); p_L is external load pressure (bar); A_c is effective area of regulator (m²).

When the external load pressure p_L increases, the regulator moves to right, the output flow of the hydraulic variable displacement pump increases, and the output pressure of hydraulic pump p_p increases, finally achieving the balance of the regulator again. At this time, the pressure differential across the throttle valve also remains constant, which is always equal to the spring preset pressure of regulator, and vice versa. This ensures that the actuator has a constant speed output and an LS control function when the external load pressure changes.

The inlet pressure of the throttle valve is equal to the output pressure of the hydraulic variable displacement pump, and the outlet pressure of the throttle valve is equal to the load pressure. The pressure differential between the output pressure of hydraulic variable displacement pump and the load pressure is equal to the spring pressure of regulator, so the pressure differential across the throttle valve is equal to the spring preset pressure of regulator.

In the actual working process, according to the equation (13-3), due to the deformation amount Δx of the regulator spring is small, $K \cdot \Delta x \ll F$, the actual spring force F is approximately equal to the preset spring force F_0. That is, the spring force F basically remains constant:

$$F = K \cdot (x_0 + \Delta x) \tag{13-3}$$

where F is preset spring force; K is stiffness coefficient of spring; Δx is deformation of spring (neglected in this experiment due to $\Delta x \ll x_0$); x_0 is initial deformation of spring.

p_p is output pressure of hydraulic pump(bar); p_L is external load pressure (bar); Δp is pressure differential across throttle valve (bar).

(2) LS Control Circuit for Double Actuators

Figure 13.2 shows the schematic diagram of the LS control circuit with dual actuators. In the actual working process, there are often two or more actuators running at the same time in the circuit, which have different loads. The shuttle valve can transmit the maximum load pressure of the system to the regulator. Therefore, the valve can maintain the pressure differential across the throttle valve on the maximum load branch to achieve load sensitivity. However, the throttle valve on the lower load branch has a pressure differential across the load that varies with the maximum load, which causes the operating condition to vary with the maximum load pressure in the system. The pressure differential across the throttle valve 2 is always equal to the preset spring pressure, but the pressure differential across the throttle valve 1 varies with the load on the actuator 2. Therefore, when the opening area of the throttle valve is constant, the running speed of the hydraulic cylinder 2 is always constant, and the running speed of the hydraulic cylinder 1 will vary with the change of the maximum load.

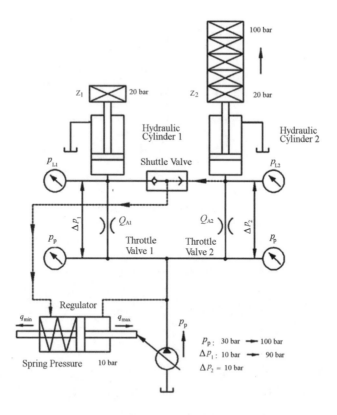

Figure 13.2 LS Circuit with Dual Actuators

(3) LS Control Circuit with Pressure Compensator

In the LS control circuit with actuators, the operating state of the actuator that is subjected to the lower load is affected by the maximum load of the system, which is not good in the actual working process. In order to solve this problem, an LS control circuit with a pressure compensator is introduced, and its working principle is shown in Figure 13.3.

The pressure compensator in the circuit is a speed regulating valve composed of a pressure reducing valve and a throttle valve. The inlet side of the throttle valve is connected to the spring-free side of the valve spool, and the outlet side of the throttle valve is connected to the spring-loaded side of the valve spool. The cross-sectional area of the valve spool of the pressure reducing valve is A_1, and the preset spring force is F_1 (the preset spring force can be set as needed). p_1 is the inlet pressure of the pressure reducing valve, p_2 is the outlet pressure of the pressure reducing valve (throttle inlet pressure), p_3 is the outlet pressure of throttle valve, and Δp_{DW} is the pressure differential across the pressure reducing valve.

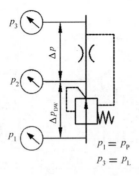

Figure 13.3   Structure of the Pressure Compensator

The force analysis on both ends of the valve spool of the pressure reducing valve can be used to obtain the following relationship:

$$F_1 = k(x_0 + x_v) \approx kx_0 = \text{constant} \tag{13-4}$$

$$p_2 \times A_1 = p_3 \times A_1 + F_1 = p_L \times A_1 + F_1 \tag{13-5}$$

$$p_2 - p_1 = \frac{F_1}{A_1} = \text{constant} \tag{13-6}$$

where F_1 is preset spring force; K is stiffness coefficient of spring; x_0 is predeformation of spring; x_v is deformation of spring (neglected in this experiment).

It can be seen from the equation (13-6) that the pressure differential $\Delta p(p_2 - p_L)$ of the throttle valve remains constant. The characteristics of the oil flow rate comply with the equation (13-1). That is to say, the operating condition of the actuator is not affected by the maximum load of the system.

As shown in Figure 13.4, in any actuator branch, the output pressure of the hydraulic variable displacement pump is the sum of the pressure differential Δp_{DW} across the pressure reducing valve, the pressure differential Δp across the throttle valve, and the load pressure p_L, which satisfies the equation (13-7).

$$p_P = \Delta p_{DW} + \Delta p + p_L \tag{13-7}$$

For circuits with multiple actuators, the pressure reducing valve in the pressure compensator cannot be effectively adjusted if the sum of the flow required by all actuators is greater than the maximum output flow of the hydraulic variable displacement pump. Although the hydraulic variable displacement pump operates at maximum flow at this time, the function of the LS system will be limited or disabled, and the system becomes a by-pass throttle speed-regulating circuit with multi-actuator. It is subjected to different load actuators.

When the LS control loop with pressure compensator is subjected to actuators with different loads, it is different from the by-pass throttle control loop. The initial valve

Lab 13 Hydraulic LS Control Experiment for Engineering Machinery

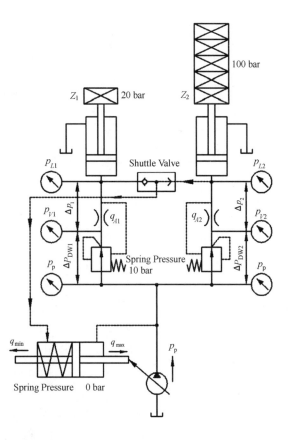

Figure 13.4 LS Control Circuit with Pressure Compensator

opening of the actuator with different loads is the same when it starts running. That is to say the "starting point" of the movement of actuator is independent of the load and flow.

(4) Cavitation Phenomenon

Because of the vertical installation of the hydraulic cylinder, the 80 kg load will make the hydraulic cylinder drop rapidly, thus forming a negative pressure in the piston side of the hydraulic cylinder. This phenomenon is called cavitation phenomenon. This type of load is called "negative load". At this time all the output oil of the hydraulic pump enters the hydraulic cylinder.

During the experiment, the manual directional spool travels over the hydraulic cylinder to lower the "starting point", and the hydraulic cylinder will drop rapidly, at which time the pump reaches the maximum output flow. Then the loader will quickly hit the spring. As the spring compresses, the loader will have a drop limit point, and the cavitation will be stopped. The spring bounces the loader up, and the loader and spring return to equilibrium quickly. After a brief pause in the equilibrium position, the oil will re-establish pressure in the piston side of the cylinder. When the pressure can push the loader down further, the loader will continue to drop until the spring is compressed to its limit, and the drop process is ended. (Cavitation phenomenon can be eliminated by installing a counterbalance valve in the oil return during hydraulic cylinder falling down.)

(5) Case Analysis-LS Circuit Operation When System Flow Is Insufficient

In engineering practice, a decrease in the engine speed of the traveling machine or an internal leakage of the control block may result in insufficient system flow. As shown in Figure 13.4, the load on the hydraulic cylinder 1 is 20 bar, and the load on the hydraulic cylinder 2 is 20~100 bar. If the demand flow of the hydraulic cylinders 1 and 2 is 100 L/min, the maximum output flow of the hydraulic pump is 100 L/min. At this time, the system flow is insufficient, and the oil preferentially flows to the hydraulic cylinder 1. Due to the spring pressure in the pressure reducing valve is 10 bar and the minimum load pressure is 20 bar, the minimum pressure at which the pressure reducing valve enters the working condition is 30 bar. When all the oil flows to the hydraulic cylinder 1, if the pressure differential generated between the two ends of the throttle valve 1 is 8 bar, the output pressure of the hydraulic variable displacement pump can only be maintained at 28 bar. At this time both pressure reducing valves in the system are inactive, and the hydraulic cylinder 2 does not move.

If the flow required by the hydraulic cylinder 1 is changed to 75 L/min, the hydraulic cylinder 2 can obtain a flow of 25 L/min. At this time, the hydraulic cylinder 2 will move at a low speed, the output pressure of the hydraulic variable displacement pump is 120 bar, and the pressure reducing valves are all in operation.

In the actual working process, the preferential flow of oil to the lower load due to insufficient system flow may adversely affect the production. Therefore, for LS systems where multiple actuators operate simultaneously, the total output flow of the hydraulic variable displacement pump needs to be sufficient at all times.

2. Structure and Working Principle of Control Block M4-12

In this experiment, the control block M4-12 of Bosch Rexroth is used. The external structural setup of M4-12 is shown in Figure 13.5. The structure and schematic principle of directional valve in M4-12 are shown in Figure 13.6 and Figure 13.7, and the technical data of M4-12 is shown in Table 13.1.

Lab 13 Hydraulic LS Control Experiment for Engineering Machinery

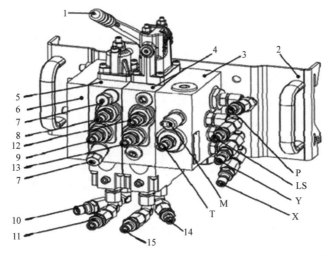

1—Control Lever; 2—Retaining Sheet; 3—Input Element; 4—Directional Valve Disc 1; 5—Directional Valve Disc 2; 6—End Element; 7—Secondary Pressure Relief Valve(Directional Valve Disc 2); 8—Working Port "B" of Directional Valve Disc 2; 9—Working Port "A" of Directional Valve Disc 2; 10—Pilot Oil Port "a" of Directional Valve Disc 2; 11—Pilot Oil Port "b" of Directional Valve Disc 2; 12—Working Port "B" of Directional Valve Disc 1; 13—Working Port "A" of Directional Valve Disc 1; 14—Pilot Oil Port "a" of Directional Valve Disc 1; 15—Pilot Oil Port "b" of Directional Valve Disc 1; P—Port for Pressure Line; LS—Load Sensing Connection; Y—Port for Pilot Oil Return; X—Port for Pilot Oil Supply; M—Manometer Connection; T—Port for Relief Line

Figure 13.5 External Structural Set-up of M4-12

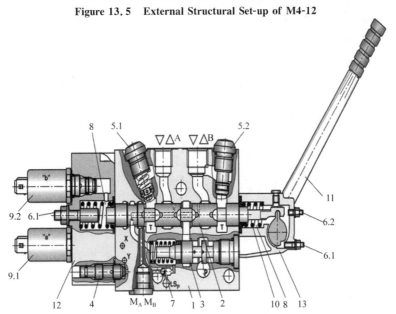

1—Valve Body; 2—Valve Element; 3—Pressure Compensator; 4—LS Overflow Valve; 5.1—Overflow Valve with Oil Filling Function; 5.2—Plug Screw; 6.1—A-end Travel Limiter; 6.2—B-end Travel Limiter; 7—LS Shuttle Valve; 8—Spring Cavity; 9.1—Pilot Control Valve "a"; 9.2—Pilot Control Valve "b"; 10—Spring; 11-Handle; 12—Oil Port A; 13—Oil Port B

Figure 13.6 Structure Principle of Directional Valve in M4-12

Control block M4-12: The directional valves are proportional valves according to the LS principle.

Actuator control: The control spool 2 is used to determine the flow direction and the flow rate that reaches the actuator ports (A or B). Pressure reducing valves 9 control the position of the control spool 2. The size of the electric current on the pressure reducing valve determines the level of the pilot pressure in the spring chambers 8 and thereby the stroke of the control spool (P → A; P → B). The pressure compensator 3 keeps the pressure differential on the control spool 2 and thereby the flow to the consumers constant.

Load pressure compensation: The pressure compensator 3 regulates pressure changes on the consumers or on the pump. The flow to the consumers remains constant, even with different loads.

Flow limitation: The maximum flow can be individually set by using the stroke limitations 6.

Pressure limitation: The LS pressure for each consumer port can optionally be overridden internally via the LS pressure relief valves 4.

Secondary pressure relief valves with large nominal size with combined oil-boost function 5 protect consumer ports A and B against pressure peaks.

The highest load pressure to the pump is transferred via the LS line and the integrated shuttle valves 7.

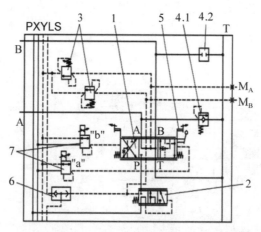

1—Valve Element; 2—Pressure Compensator; 3—LS Overflow Valve; 4.1—Overflow Valve with Oil Filling Function; 4.2—Plug Screw; 5—A-end Travel Limiter and B-end Travel Limiter; 6—LS Shuttle Valve; 7—Pilot Control Valve; P—Pump Oil Port; A,B—Oil Port of Actuator; T—Oil Return Port; X—Source Port of Control Oil; Y—Return Port of Control Oil; LS—Load Sensitive Oil Port; M_A, M_B—External LS Oil Port

Figure 13.7 Schematic Diagram of Directional Valves M4-12

Note the difference between control blocks SM-12 and M4-12.

SM-12 is a control block used in the 6-way throttle control circuit. Its principle is to control the by-pass throttle circuit of the actuator through the movement of the spool on the directional valve, so as to control the flow and pressure on the main oil circuit. When there are multiple actuators in the circuit, the operation of the actuator is affected by the maximum load of the system.

M4-12 is the control block used in LS control circuit. Its principle is to adjust the flow of pump according to the maximum load pressure of the system, and then use pressure compensator to realize that the flow of the actuators is proportional to the handle deflection degree, so that the operation of the actuators is not affected by the maximum load of the system. However, when the system flow is insufficient, the flow will preferentially flow to the actuator with the lower load, while the actuator that is subjected to the higher load can only be supplied with the remaining oil or stop running.

Table 13.1 Technical Data of M4-12

Denomination	Unit	Value
Max. Pressure	bar	120
Integrated Pilot Oil Supply:		
Control Pressure	bar	35
Max. Pilot Pressure	bar	12
1st Directional Valve:		
Secondary Pressure Limitation at A; at B	bar	70
Flow from P-A; from P-B Reduction via Stroke Limitation	L/min	5 Maximum Value upon Delivery
LS Pressure Relief Valve at A; at B	bar	60
Operating Temperature	℃	20~60
Manometer Connections	Threaded Coupling AB20-11/K1 G 1/4	

Preparation

The workbench adopts the training station of Bosch Rexroth WS290. The schematic diagram of hydraulic system circuit is shown in Figure 13.8. And the diagram of component connection in the WS290 workbench is shown in Figure 13.9. The list of hydraulic components required for the experiment is shown in Table 13.2.

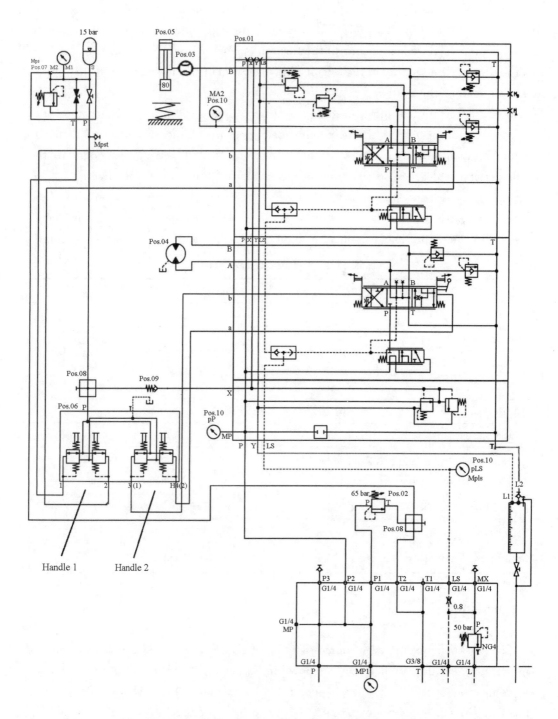

Figure 13.8 Schematic Diagram of the Hydraulic LS Control System

Lab 13 Hydraulic LS Control Experiment for Engineering Machinery

Table 13.2 **List of Components**

Position	Number	Component Name	Model
01	1	Traveling Machinery Control Block	2M4-12B
02	1	Pressure Relief Valve	DD1.1
03	1	Flow-meter	DZ30
04	1	Hydraulic Motor	DM8
05	1	Double-acting Cylinder with Single-sided Piston Rod with Load	ZY
06	1	Handle	2-2TH6
07	1	Accumulator Safety Block for Diaphragm-type Accumulator	DZ3.2
08	4	Distributor Plate with Four Ports	DZ4.2
09	1	Check Valve	DS2.1
10	3	Pressure Gauge with Hose and Quick Release Coupling without Check Valve	DZ1.4
11	Several	Hose	VSK 1
12	Several	Hose with Quick Release Coupling with Check Valve	DZ25.1

Figure 13.9 **Diagram of Component Connection in the WS290 Workbench**

Procedure

(1) Connect the circuit correctly on the workbench WS290 according to Figure 13.8 and Figure 13.9, and then check it.

Note the operating requirements of LS pump station.

The LS port of the control block must be connected to the system LS port. If there is no pressure at oil port LS, the outlet pressure p of hydraulic pump is the same as the spring force of displacement regulator, about 15 bar, which cannot provide the pressure required for the normal use of the system.

(2) Open the workbench. Insert the power plug, turn on the switch button of workbench master, turn on the switch button of hydraulic pump, and rotate the 3-way ball valve to the working position (handle level). Then the hydraulic oil is normally supplied, and the workbench is ready to operate.

(3) Supply the control oil. The accumulator is charged by closing the manual relief (hand wheel) at Pos. 09, and the pressure relief valve DD2 at Pos. 11 is set to 35 bar.

(4) Operate singly. In order to calibrate the moving direction of the actuator, handles 1 and 2 are operated separately to make the hydraulic cylinder and hydraulic motor run in different directions. Observe the up and down of the hydraulic cylinder and the clockwise and anti-clockwise rotation of the hydraulic motor. The outlet pressure of the hydraulic pump and the pressure of the LS oil line are recorded in lab report Table R.13.1 respectively (three groups of each record).

(5) Operate parallelly 1. Firstly, handle 2 is operated to make the hydraulic motor run clockwise at 250 r/min, keeping the deflection angle of handle 2 constant. Then operate the handle 1 to slowly raise the hydraulic cylinder, observe the change of the hydraulic motor speed during this process, and record the speed of the hydraulic motor in lab report Table R.13.2 when the flow through the hydraulic cylinder is changed.

(6) Operate parallelly 2. First, handle 1 is operated to make the hydraulic cylinder rise slowly, keeping the deflection angle of handle 1 constant. Meanwhile, operate handle 2 to rotate the hydraulic motor clockwise to the maximum speed gradually, and then reverse the motor to the maximum speed. Observe the change in the rising speed of the hydraulic cylinder during this process.

(7) Shut down the workbench. Put the actuator in the no-load state, rotate the 3-way ball valve to the closed position (handle vertical), control the directional valve spool for several times to remove the remaining pressure in the control block, power off the hydraulic pump, open the manual relief (hand wheel) to discharge the accumulator, and turn off the master switch button of the test bed and pull out the power plug.

(8) Clean the workbench and the experiment is finished.

Note: Because the LS control system requires a large amount of flow, there may be insufficient flow in the experiment. To illustrate the principle of an LS system, during the experiment, we can operate the deflection angle of the handle to control the flow through the actuator to demonstrate the function of the LS system when the flow is sufficient and insufficient and then answer the questions 3 and 4 in the lab report.

Lab 14 LUDV[①] Control Experiment for Engineering Machinery

Objectives

(1) Recognize the hydraulic components in engineering machinery, and learn to correctly connect the hydraulic LUDV control circuit;

(2) Understand the working principle of hydraulic LUDV control circuit of engineering machinery;

(3) Understand the performance of the actuators in the LUDV control system under the conditions of sufficient flow and insufficient flow.

Equipment Preparation

This experiment makes use of WS290 from Bosch Rexroth of Germany. Detail information is seen in Figure 12.1 of lab 12.

Principle

1. Working Principle of LUDV Control Circuit

(1) LUDV Control Circuit for a Single Actuator

The LUDV working principle is shown in Figure 14.1. At this point, the pressure compensator does not work, its valve port is fully open, and the circuit is equivalent to the load sensing circuit. The outlet pressure p_N of the throttle valve is equal to the pressure p_{LS} of the signal oil line. And the pressure difference Δp across the throttle valve is equal to the preset pressure of spring in the displacement regulator of the hydraulic pump (hereinafter referred to as the regulator), and the preset pressure can be adjusted. Assuming that the preset pressure of the spring in this circuit is 10 bar, the pressure difference Δp across the throttle valve is as shown in the following equation:

[①] Load Independent Flow Distribution is the same meaning as LUDV. LUDV is originally drawn from the Germany, and it is an abbreviation of Germany.

Lab 14 LUDV Control Experiment for Engineering Machinery

$$\Delta p = p_p - p_N = p_p - p_{LS} = 10 \text{ bar} \tag{14-1}$$

where p_p is output pressure of hydraulic pump (bar); q_p is output flow of hydraulic pump (L/min); p_{LS} is pressure of LS signal oil line (bar); p_N is outlet pressure of throttle valve (bar); p_L is external load pressure (bar); Δp is pressure differential across throttle valve (bar); Δp_D is pressure difference across the pressure compensator (bar).

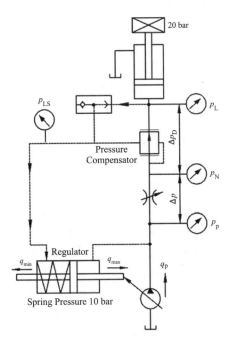

Figure 14.1 LUDV Control Circuit for a Single Actuator

The oil flow rate through the throttle valve can be calculated according to the flow equation of the thin-wall orifice. According to the following equation, when the opening of the throttle valve is constant and the pressure differential is constant, the flow through the throttle valve is constant. That is, the operation of the actuator is not affected by the maximum load pressure of the system.

$$q = CA \sqrt{\frac{2\Delta p}{\rho}} \tag{14-2}$$

where q is flow rate through the orifice (L/min); C is discharge coefficient (C is 0.62 in this experiment); $A = \dfrac{\pi d^2}{4}$ is section area of orifice (mm^2); d is throttle valve opening (mm); Δp is pressure differential at both ends of orifice (bar); ρ is oil density (kg/m^3).

In order to facilitate the engineering data recording, $q, A, d, \Delta p$ make use of engineering units in this lab.

(2) LUDV Control Circuit for Double Actuators

In the actual working process, there are often two or more actuators running at the same time in the circuit, and the load pressure is different. The shuttle valve can transmit the maximum load pressure of the system to the regulator. Figures 14.2 and 14.3 show the LUDV control loop of the double actuators when the hydraulic pump flow is sufficient and insufficient. At this time, the pressure compensator starts to work. The pressure p_{LS} of the signal oil passage is always equal to the outlet pressure p_N of the throttle valve. Then, according to the relationship between the maximum load pressure and the load pressure of the branch, the pressure is adjusted to keep the pressure difference Δp at both ends of the throttle valve constant, so that the stable operation of the hydraulic system can be realized. The following is the introduction of LUDV control loops in both cases.

① $q_{p_{max}} \geqslant \sum_{1}^{n} q_i$ (q_i - Flow Required for the Actuator)

When the total flow of the hydraulic pump is greater than (or equal to) the sum of the flow rates required by all actuators ("Required flow rate" means the flow rate determined by the open area of the throttle valve at a constant pressure differential), the output pressure of the hydraulic pump is the sum of the maximum load pressure and the spring preset pressure of the regulator. At this point, for the two actuators, the valve port of the pressure compensator in the branch under higher load is fully open, and the pressure difference across the pressure compensator is 0 bar approximately. The LS signal makes the pressure compensator port of the branch under lower load become small until its inlet pressure is equal to the load sensing pressure. At this time, the pressure difference across the pressure compensator is the difference between the two load pressures. By this principle, the pressure difference Δp across the throttle valve on both branches is equal to the preset pressure of the regulator spring, so the flow obtained by each actuator is determined only by the opening area of the throttle valves and independent of load pressure.

Example 1: Assume that the maximum output flow of the hydraulic pump is 100 L/min and the spring preset pressure of the regulator is 10 bar. The flow required for actuator 1 is $q_{A1} = 50$ L/min (Indicate the flow rate is determined when the pressure difference across the throttle valve is 10 bar and the opening area is A1.) and the load pressure is 20 bar. The flow rate required for actuator 2 is $q_{A2} = 50$ L/min (Indicate the flow rate is determined when the pressure difference across the throttle valve is 10 bar and the opening area is A2.) and the load pressure is 100 bar. The output pressure of the hydraulic pump is $100+10=110$ bar, and the output flow of the hydraulic pump is $50+50=100$ L/min, as shown in Figure 14.2. The pressure compensator that is subjected to the higher pressure load branch is fully open, and the pressure difference between the inlet and outlet is 0 bar. The pressure compensator that bears the lower pressure load branch is semi-closed, and the pressure difference between the two ends is $100-20=80$ bar. The pressure difference across the throttle valves is 10 bar, and the flow rate obtained on each actuator is only proportional to the opening area of the throttle valves.

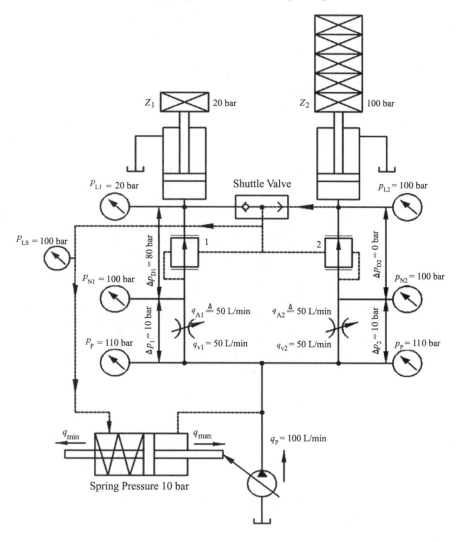

Figure 14.2 LUDV Control Loop for Double Actuators When System Flow Is Sufficient

② $q_{p_{max}} < \sum_{1}^{n} q_i$

When the total flow of the hydraulic pump is less than the sum of the flow rates required by all actuators as Figure 14.3 shows, first, the hydraulic pump will output the maximum flow, which will be distributed to all actuators according to the proportional coefficient K_q [as shown in equation (14.3)]. The output pressure of the hydraulic pump is the sum of the maximum load pressure of the system and the pressure differential across the throttle valve. At this point, for the two actuators, the pressure compensator in the branch under higher load is fully open, and the pressure differential across the pressure compensator is 0 bar approximately. The LS signal makes the opening area of the pressure compensator of the branch under lower load become small until its inlet pressure is equal to the load sensitive pressure. At this time, the pressure differential across the pressure compensator is the difference between the two load pressures. According to the actual flow rate obtained by the actuator, the pressure difference across the throttle valve can be calculated. Similarly, the flow rate obtained by the two actuators also depends only on the opening area of the throttle valve, regardless of the magnitude of the load pressure.

$$K_q = \frac{\sum_1^2 q_i}{q_{P_{max}}} \qquad (14\text{-}3)$$

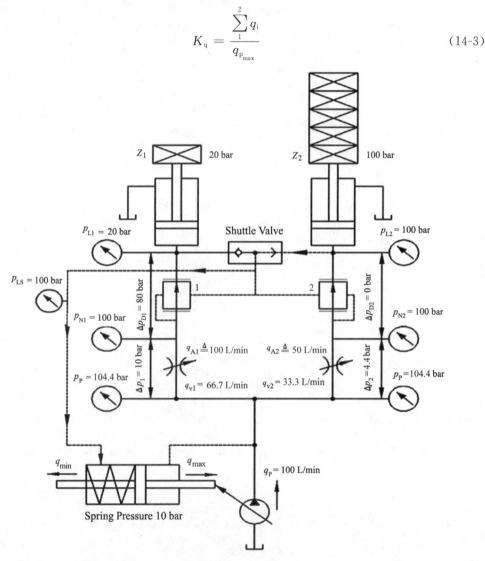

Figure 14.3 LUDV Control Loop for Double Actuators When System Flow Is Insufficient

Example 2: Assume that the maximum output flow of the hydraulic pump is 100 L/min and the spring preset pressure of the regulator is 10 bar. The flow required for actuator 1 is

$q_{A1}=100$ L/min and the load pressure is 20 bar. The flow rate required for actuator 2 is $q_{A2}=50$ L/min and the load pressure is 100 bar, as shown in Figure 14.3. The hydraulic pump will output a maximum flow of 100 L/min, and the ratio of the total flow required by actuators to the maximum flow of the hydraulic pump is $150/100=1.5$. In this ratio, the actuator that is subjected to the high-pressure load will get a flow of $100/1.5=66.7$ L/min, and the actuator that is subjected to the low-pressure load will get a flow of $50/1.5=33.3$ L/min. The pressure compensator in the branch under higher load is fully open, and the pressure differential across the pressure compensator is 0 bar approximately. The opening area of the pressure compensator in the branch under lower load is reduced and the pressure differential is $100-20=80$ bar. The pressure differential across the throttle valve is 4.4 bar, and the calculation method is as follows.

In the hydraulic LUDV system, the pressure difference across the throttle valve on each branch is always equal, and the flow characteristics are always the same, so two throttle valves can be considered as a single throttle valve for calculation. Using this principle, the pressure difference across the throttle valves in example 2 is calculated. The relationship between the output flow of the hydraulic pump and the opening area of the throttle valve and the pressure difference across the throttle valve is calculated by:

$$q_1 = CA_1\sqrt{\frac{2\Delta p_1}{\rho}} \tag{14-4}$$

$$q_2 = CA_2\sqrt{\frac{2\Delta p_2}{\rho}} \tag{14-5}$$

$$\frac{\Delta p_1}{\Delta p_2} = \left(\frac{A_2}{A_1}\right)^2 \tag{14-6}$$

$$\Delta p_2 = \Delta p_1 \times \left(\frac{A_1}{A_2}\right)^2 = 10 \times \left(\frac{2}{3}\right)^2 = 4.4 \text{ bar} \tag{14-7}$$

where q_1 and q_2 are the hydraulic pump output flow rates in examples 1 and 2, both of which are 100 L/min. The sum of the flow rates required for each of the actuators in example 1 is 100 L/min, and the sum of the flow rates required for each of the actuators in example 2 is 150 L/min. The ratio of the opening areas of the equivalent throttle valves in examples 1 and 2 is $100/150=2/3$. The ratio of the pressure difference between the two ends of the equivalent throttle valve in examples 1 and 2 is shown in the equation (14-6). The pressure difference between the two ends of the equivalent throttle valve in examples 2 is shown in the equation (14-7). Therefore, the pressure difference across the throttle valves in example 2 is $\Delta p_2 = 4.4$ bar.

2. Structure and Working Principle of Control Block SX-12

In this experiment, the valve SX-12 of Bosch Rexroth is used. The external structural set-up of SX-12 is shown in Figure 14.4, the structure and working principle of directional valve in SX-12 are shown in Figure 14.5 and Figure 14.6, and the technical data of SX-12 is shown in Table 14.1.

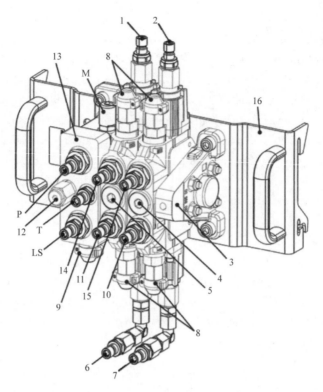

1—Pilot Oil Port "b" Directional Valve Disc 1; 2—Pilot Oil Port "b" Directional Valve Disc 2;
3—End Element; 4—Directional Valve Disc 2; 5—Directional Valve Disc 1; 6—Pilot Oil Port "a"
Directional Valve Disc 2; 7—Pilot Oil Port "a" Directional Valve Disc 1; 8—Sealed Secondary
Pressure Relief Valve; 9—LS Pressure Relief Valve; 10—Working Port "A" Directional Valve
Disc 2; 11—Working Port "A" Directional Valve Disc 1; 12—LS Flow Control Valve; 13—Input
Element; 14—Working Port "B" Directional Valve Disc 1; 15—Working Port "B" Directional
Valve Disc 2; 16—Retaining Sheet; M—Manometer Connection; P—Port for Pressure Line;
T—Port for Relief Line; LS—Load Sensing Connection

Figure 14.4 External Structure Set-up of SX-12

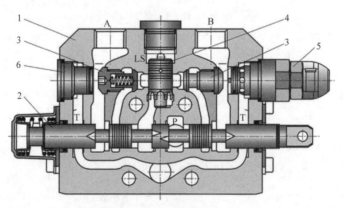

1—Valve Body; 2—Valve Spool; 3—Check Valve; 4—Pressure Compensator;
5—Second Valve; 6—Plug Screw

Figure 14.5 Structure Principle of Directional Valve in SX-12

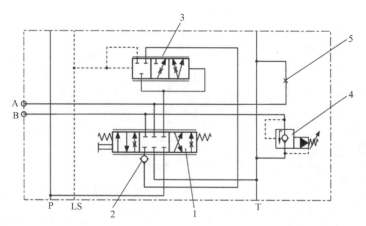

1—Valve Spool; 2—Check Valve; 3—Pressure Compensator; 4—Second Valve;
5—Plug Screw; P—Pump; A,B—Actuators; T—Tank; LS—Load-sensing

Figure 14.6 Schematic Diagram of Directional Valves SX-12

Note the difference between control blocks M4-12 and SX-12.

M4-12 is the control block used in LS control circuit. From the order of the oil flow direction inside the control block, the pressure compensator is located before the throttle valve and this control mode is called "meter-in" control system. The principle of the pressure compensator in the control block is a pressure reducing valve. SX-12 is the control block used in LUDV control circuit. From the oil flow direction inside the control block, the pressure compensator is located after the throttle valve and this control mode is called "meter-out" control system.

When the system flow is sufficient, M4-12 and SX-12 can maintain the constant pressure difference across the throttle valve through pressure compensation, so that the operation of each actuator is not affected by other loads. However, when the system flow is insufficient, the M4-12 will preferentially supply the actuators that are subjected to the lower load, while the actuators that are subjected to the higher load can only be supplied with the remaining oil or stop running. At this point, the load-sensitive control loop becomes a by-pass throttle control loop. The SX-12 distributes the oil to the actuators according to the opening ratio of each branch throttle valve. Each actuator will continue to run stably at a lower speed.

Table 14.1 Technical Parameters of SX-12

Denomination	Unit	Value
Max. Pressure	bar	120
Integrated Pilot Oil Supply:		
Control Pressure	bar	35
Max. Pilot Pressure	bar	12

(continued)

Denomination	Unit	Value
1st Directional Valve:		
Secondary Pressure Limitation at A; at B	bar	70
Flow from P-A; from P-B Reduction via Stroke Limitation	L/min	5 Maximum Valve upon Delivery
LS Pressure Relief Valve at A; at B	bar	60
Operating Temperature	℃	20~60
Hydraulic Connections	Quick Release Coupling with Connector, Type W	
Manometer Connections	Threaded Coupling AB20-11/K1 G 1/4	

Preparation

The workbench adopts Bosch Rexroth WS290 workbench. The diagram of hydraulic system circuit is shown in Figure 14.7. And the diagram of components connection in the WS290 workbench is shown in Figure 14.8. The list of hydraulic components required for the experiment is shown in Table 14.2.

Solution to the Cavitation Phenomenon

Because of the vertical installation of the hydraulic cylinder, the hydraulic system will have a cavitation phenomenon during the descent of the cylinder. For this phenomenon, a counterbalance valve can be installed in the oil return pipe line during cylinder fall to solve this problem. The working principle of the counterbalance valve is shown in Pos. 09 in Figure 14.8. When the cylinder rises, the oil directly passes through the check valve, and the oil line flows normally. When the cylinder falls down, the counterbalance valve starts to work, and causes back pressure on the returning oil line, so that the downward stroke of the cylinder can be stabilized and the cavitation phenomenon can be eliminated.

Lab 14 LUDV Control Experiment for Engineering Machinery

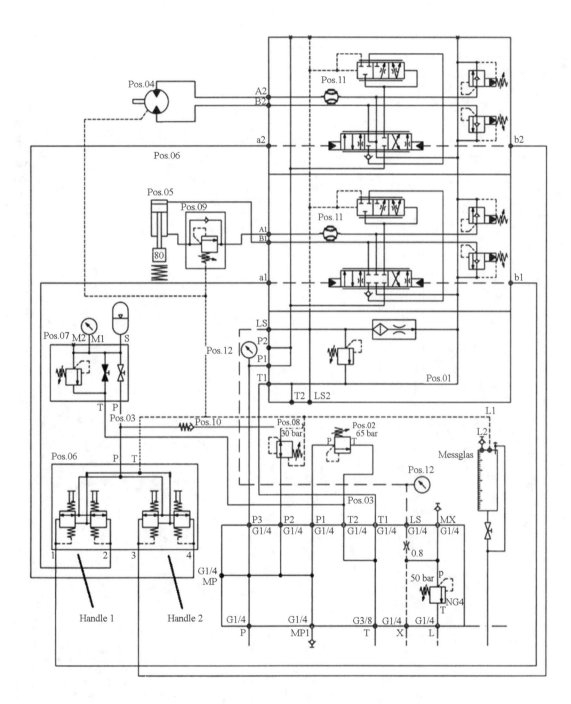

Figure 14.7 Schematic Diagram of LUDV Control System

Table 14.2 List of Components

Position	Number	Component Name	Model
01	1	Traveling Machinery Multi-way Valve	2SX-12
02	1	Pressure Relief Valve	DD1.1
03	2	Distributor Plate with Four Ports	DZ4.2
04	1	Hydraulic Motor	DM8
05	1	Double-acting Cylinder with Single-rod Piston Rod with Load	ZY
06	1	Handle	2-2TH6
07	1	Accumulator Safety Block for Diaphragm-type Accumulator	DZ3.2
08	1	Reducing Valve	DD2.N
09	1	One-way Relief Valve	DD3
10	1	Check Valve	DS2.1
11	2	Flow-meter	DZ30
12	2	Pressure Gauge with Hose and Quick Release Coupling without Check Valve	DZ1.4
13	Several	Hose	VSK 1
14	Several	Hose with Quick Release Coupling with Check Valve	DZ25.1

Figure 14.8 Diagram of Components Connection in the WS290 Workbench

Procedure

(1) Connect the circuit correctly on the workbench WS290 according to Figure 14.7 and Figure 14.8, and then check it.

Note the operating requirements of LS pump station(LS:Load Sensing).

The LS port of the control block must be connected to the system LS port. If there is no pressure at oil port LS, the outlet pressure p of hydraulic pump is the same as the spring force of displacement regulator, about 15 bar, which cannot provide the pressure required for the normal use of the system.

(2) Start up the workbench. Insert the power plug, turn on the switch button of workbench master, turn on the switch button of hydraulic pump, and rotate the ball valve of three-position to the working position (handle level). Then the hydraulic oil is normally supplied, and the workbench is ready to operate.

(3) Supply the control oil. The accumulator is charged by closing the manual relief (hand wheel) at Pos. 09, and the pressure relief valve DD2 at Pos. 11 is set to 35 bar.

(4) Operate singly. In order to calibrate the moving direction of the actuator, handles 1 and 2 are operated separately to make the hydraulic cylinder and hydraulic motor run in different directions. Observe the up and down of the hydraulic cylinder and the clockwise and counter clockwise rotation direction of the hydraulic motor, and record the flow values of the hydraulic cylinder and the hydraulic motor at the maximum deflection of the handle, and fill in the lab report.

(5) Operate parallelly 1. First of all, handle 2 is operated to make the hydraulic motor at the maximum speed, keeping the deflection angle of handle 2 constant. Then, the handle 1 is operated to raise or lower the hydraulic cylinder, and the amount of deflection is gradually increased from small to large to observe the change of the rotational speed of the hydraulic motor during the process. When both the handle 1 and the handle 2 are at the maximum deflection amount, record the flow values of the hydraulic cylinder and the hydraulic motor in the lab report.

(6) Operate parallelly 2. First, handle 1 is operated to make the hydraulic cylinder rise slowly, keeping the deflection angle of handle 1 constant. At the same time, handle 2 is operated to make the hydraulic motor rotate at maximum speed. Observe the change in the rising speed of the hydraulic cylinder during this process. When the hydraulic motor rotates clockwise at different speeds, record the flow rates through the hydraulic cylinder in lab report Table R. 14.1.

(7) Shut down the workbench. Put the actuator in the no-load state, rotate the ball valve of three-position to the closed position (handle vertical), control the directional valve spool for several times to remove the remaining pressure in the control block,

power off the hydraulic pump, open the manual relief (hand wheel) to discharge the accumulator, and turn off the master switch button of the test bed and pull out the power plug.

(8)Clean the workbench and the experiment is finished.

Note: Because the LUDV control system requires a large amount of flow, there may be insufficient flow in the experiment. To illustrate the principle of an LUDV system, during the experiment, the change in the deflection angle of the handle can be used to control the flow through the actuator to demonstrate the function of the LUDV system when the flow is sufficient and insufficient.

Lab Report

Lab Report 1 Reynolds Experiment

Name _____ Study ID _____
Test Stand _____ #. Date _____

Results and Calculations

Pipe Diameter $D = 13 \times 10^{-3}$ m, Area $A = 133 \times 10^{-6}$ m^2

Table R.1.1 Flow Rate Record

Temperature θ/℃	Volume V/m^3	Time t/s		
		t_1	t_2	t_3
0				
10				
20				
30				
40				
50				
60				

Calculations for Each Setting

Flow rate $q = V/t$ (m³/s) =
Velocity $v = q/A$ (m/s) =
Kinematic viscosity $\nu =$ (see Figure 1.1)
Reynolds number $Re = \dfrac{vD}{\nu} =$

$v_{\text{crit}} = \dfrac{q_{\text{crit}}}{A} =$ (by observation of decreasing velocities)

$v^{\text{crit}} = \dfrac{q^{\text{crit}}}{A} =$ (by observation of increasing velocities)

(1) Which factors affect the flow state (laminar/turbulent/transient)? Please state them according to the experiment.

(2) Is v_{crit} equal to v^{crit}? Why?

Lab Report

Lab Report 2　Bernoulli's Experiment

Name _____　　Study ID _____

Test Stand _____ #.　　Date _____

Table R.2.1　Experiment Records

Volume Collected V/m^3	Time to Collect t /s	Flow Rate q_v /(m^3·s^{-1})	h	Distance into Duct /m	Area of Duct A /10^{-6}m^2	Static Head h /m	Velocity v /(m·s^{-1})	Dynamic Head h_d /m	Total Head h_t /m
			h_1	0.00	490.9				
			h_2	0.060 3	151.7				
			h_3	0.068 7	109.4				
Average flow rate			h_4	0.073 2	89.9				
			h_5	0.081 1	78.5				
			h_6	0.141 5	490.9				

The above table should be constructed for each set of readings:

Application of Theory

(1) Comment on the validity of the Bernoulli equation for convergent flow and divergent flow.

(2) State clearly the assumptions made in deriving the Bernoulli equation and justification for all your comments.

(3) Comment on the comparison of the total heads obtained by the two methods you have carried out.

Lab Report 3 Fluid Friction

Name _____   Study ID _____
Test Stand _____ #.   Date _____

Part 1 Fluid Friction in a Smooth Bore Pipe Results and Analysis

Table R.3.1 Fluid Friction in a Smooth Bore Pipe Calculation Formula

No.	Volume V/m^3	Time t/s	Flow Rate $q/(\mathrm{m}^3\cdot\mathrm{s}^{-1})$	Pipe Diam d/m	Velocity $u/(\mathrm{m}\cdot\mathrm{s}^{-1})$	Reynolds Number Re	λ	Calculated Head Loss $h_c/\mathrm{mH_2O}$	Measured Head Loss $h/\mathrm{mH_2O}$
1									
2									
3									
4									

Note: $q=\dfrac{V\times 10^{-3}}{t}$, $u=\dfrac{4q}{\pi d^2}$, $Re=\dfrac{\rho u d}{\mu}$, λ(From Moody Diagram), $h_c=\dfrac{\lambda L u^2}{2gd}$, $h=h_1-h_2$.

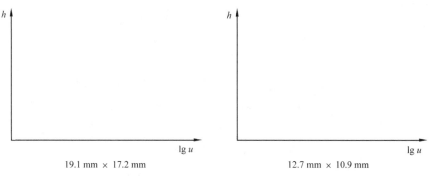

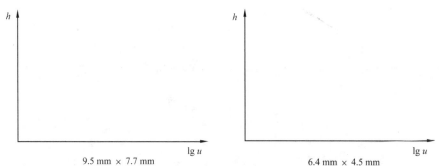

Figure R.3.1 Graphs of Fluid Friction in a Smooth Bore Pipe

Part 2 Head Loss due to Pipe Fillings Results and Analysis

Table R.3.2 Head Loss due to Pipe Fittings Calculation Formula

No.	Volume V/m^3	Time t/s	Flow Rate $q/$ $(\text{m}^3 \cdot \text{s}^{-1})$	Pipe Diam d/m	Velocity $u/$ $(\text{m} \cdot \text{s}^{-1})$	Velocity Head $h_v/\text{mH}_2\text{O}$	Measured Head Loss $h/\text{mH}_2\text{O}$	Fitting Factor K	Valve Position
1									
2									
3									
4									

Note: $q = \dfrac{V \times 10^{-3}}{t}$, $u = \dfrac{4q}{\pi d^2}$, $h_v = \dfrac{u^2}{2g}$, $h = h_1 - h_2$, $K = \dfrac{h}{h_v}$.

Question: what's n? $n = $ _____ .

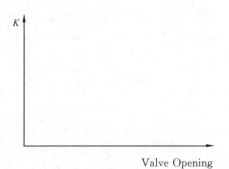

Figure R.3.2 Graph of K Factor Against Valve Opening for Each Test Valve

Part 3 Fluid Friction in a Roughened Pipe Results and Analysis

Table R. 3. 3 Calculation Formula

No.	Volume V/m^3	Time t/s	Flow Rate $q/(\text{m}^3 \cdot \text{s}^{-1})$	Pipe Diam d/m	Velocity $u/(\text{m} \cdot \text{s}^{-1})$	Reynolds Number Re	Measured Head Loss $h/\text{mH}_2\text{O}$	Friction Coefficient f
1								
2								
3								
4								

Note: $q = \dfrac{V \times 10^{-3}}{t}$, $u = \dfrac{4q}{\pi d^2}$, $Re = \dfrac{\rho u d}{\mu}$, $h = h_1 - h_2$, $f = \dfrac{gdh}{2lu^2}$.

Pipe length $L =$ _____ m.

Roughness height $K =$ _____ μm.

Plot a graph of pipe friction coefficient versus Reynolds number (log scale).

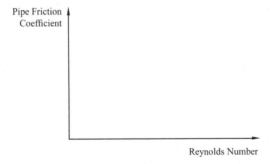

Figure R. 3. 3 Fluid Friction in a Roughened Pipe Graph of Pipe Friction Coefficient Versus Reynolds Number

Question: Note the difference from the smooth pipe curve on the Moody diagram when the flow is turbulent.

Part 4 Fluid Friction in an Orifice Plate or Venturi Results and Analysis

Table R. 3. 4 Calculation Formula for the Venturi and Orifice Plate

No.	Volume V/m^3	Time t/s	Flow Rate $q_m/(\text{m}^3 \cdot \text{s}^{-1})$	Differential Head $h/\text{mH}_2\text{O}$	Flow Rate Calculated $q_c/(\text{m}^3 \cdot \text{s}^{-1})$	Measured Head Loss $h/\text{mH}_2\text{O}$
1						
2						
3						
4						

Note: $q_m = \dfrac{V \times 10^{-3}}{t}$, $q_c = Eq3\text{-}6$, $h = h_1 - h_2$.

All readings should be tabulated as follows:

pipe lengh $L=$ _____ m;

roughness height $K=$ _____ μm.

Plot a graph of pipe friction coefficient versus Reynolds number(log scale).

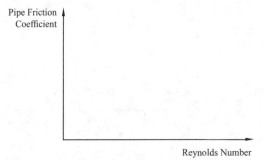

Figure R. 3. 4 Venturi and Orifice Plate Graph of Pipe Friction

Question: Note the difference from the smooth pipe curve on the Moody diagram when the flow is turbulent.

Question: Compare each calculated flow rate with the actual flow rate measured.

Lab Report

Question: Compare the head loss across the Venturi and orifice at the same flow rate.

Question: Compare the differential head across the Venturi and orifice plate at the same flow rate.

Question: Comment on the differences in the two devices and their suitability for flow measurement.

Use the theory covered by the experiment to determine the K factor for the two flow meters.

Table R.3.5 Pitot Tube Calculation Formula

Pitot Tube Position	Volume V/m^3	Time t/s	Flow Rate $q/(\text{m}^3 \cdot \text{s}^{-1})$	Pipe Diameter d/m	Pipe Area A/m^3	Velocity Measured $v_m/(\text{m} \cdot \text{s}^{-1})$	Differential Head $h/\text{mH}_2\text{O}$	Velocity Calculated $v_c/(\text{m} \cdot \text{s}^{-1})$

Note: $q = \dfrac{V \times 10^{-3}}{t}$, $A = \dfrac{\pi \times d^2}{4}$, $v_m = \dfrac{q}{A}$, $h = h_c - h_o$, $v_c = (2gh)0.5$.

Question: What is the effect of the velocity profile on the results obtained in Part 4?

Compare each calculated velocity with the measured velocity (determined from the volume flowrate and cross sectional area of the pipe).

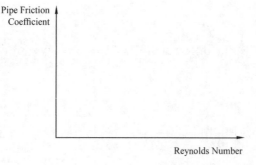

Figure R. 3. 5 Pitot Tube Graph of Pipe Friction Coefficient Versus Reynolds Number

Note: The Pitot tube is included for the purpose of demonstration only. The small differential head produced by the Pitot tube means that it should only be used in applications where high velocity is to be measured. Accuracy of measurement on the C6-ML II - 10 will be poor because of the low water velocity.

Lab Report 4 Fluid Power Components and Circuits

Name _____ Study ID _____

Test Stand _____ #. Date _____

Table R. 4. 1 **Test Stand Components**

Tag#	Component Name	Manufacturer	Model Number
1	Electric Motor		
2	Hydraulic Pump		
3	Directional Control Valve		
4	Pressure Relief Valve		
5	Return Line Filter		
6	Hydraulic Cylinder		
7	Hydraulic Motor		
8	Flow Control Valve		
9	Hydraulic Reservoir		
10	Pressure Gauge		
11	By-pass Valve		

Table R.4.2 Descriptions of Symbols (Filling the letter according to Table 4.2)

_____ Mechanical Operator

_____ Variable Displacement, Bi-directional Motor

_____ Directional Control Valve, 3-position, 4-way, Solenoid Operated with Manual Overrides, Closed Center Spool.

_____ Check Valve

_____ Temperature Gauge

_____ Strainer

_____ Spring Control

_____ Push Button Operator

_____ Directional Control Valve, Two Positions, 3-Way, Normally Passing, Solenoid Controlled

_____ Accumulator, Weight Loaded

_____ Fixed Orifice

_____ Bi-directional Fixed Displacement Pump

_____ Pressure Relief Valve

_____ Accumulator, Gas Charged

_____ Pressure Sequence Valve

_____ Adjustable Non-compensated Flow Control Valve with By-pass

_____ Manual Shutoff Valve

_____ Flow Meter

_____ Hydraulic Pilot Operator

_____ Pilot Operated Check Valve

_____ Heat Exchanger, Liquid Medium

_____ Pressure Reducing Valve

_____ Pressure Gauge

_____ Hydraulic Cylinder

Lab Report

Circuit A: Draw the complete cylinder circuit schematic in lead pencil with a straight edge and template.

Instructor signature Automation Studio animation. _____

Circuit B: Draw the complete hydraulic motor circuit schematic in lead pencil with a straight edge and template.

Instructor signature Automation Studio animation. _____

Lab Report 5   Pump Disassembly and Performance

Name _____   Study ID _____
Test Stand _____ #.   Date _____

Part 1   Pump Disassembly and Performance

Gear Pump Specifications

Manufacturer	
Model Numbe	
Shaft Rotation (Viewed from Shaft End, Clockwise or Anticlockwise Direction)	

Gear Pump Questions

Rotate the pump shaft in the direction indicated by the arrow stamped on the side of the pump housing. Describe the oil flow path from inlet port to outlet port. Sketching picture of the pump may assist in describing the flow path.

Vane Pump Specifications

Manufacturer	
Model Number	
Pressure	
Shaft Rotation (Viewed from Shaft End, Clockwise or Anticlockwise Direction)	

Vane Pump Questions

(1) How is the pump's operation changed by flipping the cam ring such that the arrow on the cam ring points in the opposite direction?

Lab Report

(2) List several areas (at least three) where clearances are critical to the proper function of this pump.

Piston Pump Specifications

Manufacturer	
Model Number	
Pressure	
Swashplate Angle Range	

Piston Pump Questions

(1) Piston pump is widely used in what occasion?

(2) How to deal with pump noise problem? Please cite 3 different ways.

Part 2 Data Analysis and Questions

Table R.5.1 Experiment Data for Procedure 2

Target Pressure/bar	Recorded Pressure/bar	Flow/(L · min^{-1})	Hydraulic Power	Volumetric Efficiency
15				
20				
25				
30				
35				
40				
45				
50				

Temperature at the Beginning _____ ℃.

Temperature at the Ending _____ ℃.

Table R.5.2 Experiment Data for Procedure 3

Target Pressure/bar	Recorded Pressure/bar	Flow/(L · min^{-1})	Hydraulic Power	Volumetric Efficiency
15				
20				
25				
30				

Lab Report

(continued)

Target Pressure/bar	Recorded Pressure/bar	Flow/(L · min⁻¹)	Hydraulic Power	Volumetric Efficiency
35				
40				
45				
50				

Temperature at the Beginning _____ ℃.
Temperature at the Ending _____ ℃.

Results

For the pump tested, attach plots presenting the following information(Please plot with different lines and labels on them):

(1) Flow versus pressure;

(2) Hydraulic power versus pressure;

(3) Volumetric efficiency versus pressure.

Plots can be generated using a spreadsheet program or drawn by hand. If the plots are drawn by hand, use fine grid paper.

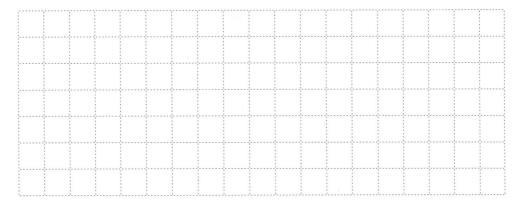

Questions

Is the simulation result coincident with the result on DS4 workbench? If not, try to find where the errors come from.

Lab Report 6 Cylinder Circuit Operation

Name _____   Study ID _____

Test Stand _____ #.   Date _____

Table R. 6. 1 Cylinder Circuit Specifications

Style		Acting
		Rod
Piston Diameter		mm
Rod Diameter		mm
Stroke Length		mm

Table R. 6. 2 Cylinder Data for Normal Connection

Opening Position of Valve DF3	Cylinder	Pressure (P_{e1}) Moving	Pressure (P_{e2}) Moving	Pressure (P_{e1}) Stop	Pressure (P_{e2}) Stop	Working Time/s	Velocity/ (mm·s^{-1})
1	Extend						
	Retract						
1.6	Extend						
	Retract						
2	Extend						
	Retract						

Table R. 6. 3 Cylinder Data for Regenerative Connection

Opening Position of Valve DF3	Cylinder	Pressure (P_{e1}) Moving	Pressure (P_{e2}) Moving	Pressure (P_{e1}) Stop	Pressure (P_{e2}) Stop	Working Time/s	Velocity/ (mm · s^{-1})
1	Extend						
	Retract						
1.6	Extend						
	Retract						
2	Extend						
	Retract						

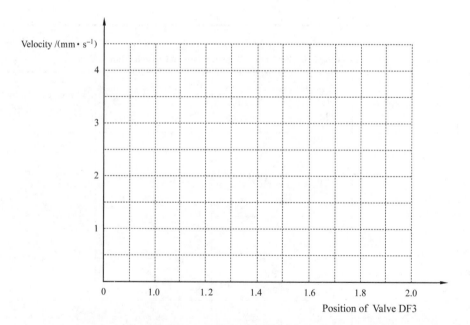

Figure R. 6. 1 Curve of Extending and Retracting of Normal and Regenerative Cylinder

Questions

(1) Compare the calculated piston extending velocity in normal mode with the calculated extending velocity in regenerative mode. How much faster does the rod extend in regenerative mode? What accounts for this velocity increase?

(2) Which is faster, regenerative extending velocity or retracting velocity in the regenerative mode? Why?

(3) Compare the extending force while in regenerative mode with the retracting force. Discuss the variation.

Lab Report 7 Hydraulic Motor

Name _____   Study ID _____

Test Stand _____ #.   Date _____

Table R. 7. 1 Normal Motor Test Stand Specifications

Electric Motor		Pump	
Power	kW	Displacement	mL/r
Speed	r/min	Pro-Rated Flow	L/min
Electric	V	Relief Pressure	Pa
Information	Phase	Driving Speed	r/min
Directional Control Valve		Motor	
Model #.		Style	Directions (One or Two)
Style	Way		
	Position		
Center Spool (Configuration)		Displacement	mL/r
Relief Valve		Flow Meter	
Pressure Setting	Pa	Maximum Flow	L/min

Table R. 7. 2 Normal Motor Test Data

Recorded			Calculated		
Pressure/Pa	Flow Rate/ (L · min⁻¹)	Speed/ (r · min⁻¹)	Torque/ (N · m)	Theoretical Flow/(L · min⁻¹)	Output Power/ kW

Calculations

Show below all related calculations required in this exercise. Write neatly and indicate what equations are used in the calculations. If more space is required, attach an additional sheet labeled "Calculations" to the end of the laboratory report.

Lab Report

Results

For each motor tested, attach plots of torque versus power. Plots can be generated using a spreadsheet program or drawn by hand. If the plots are drawn by hand, use fine grid paper.

Lab Report 8 Valve Disassembly

Name _____ Study ID _____
Test Stand _____ #. Date _____

Questions

Part 1 Directional Control Valves

(1) Obtain a directional control valve to disassemble.

(2) Complete the following list of specifications for the DCV.

    Manufacturer: _____
    Model Number: _____
    Serial Number: _____

Description:

    _____ control valve _____ position
    _____ way _____ operated

The spool is (check one):

    ☐ spring centered
    ☐ spring offset

(3) Record the following preliminary observations by using the catalog information provided in the laboratory.

    Mounting Style (Pipe, Subplate, or Flange) _____
    Center Spool Configureuration _____
    Maximum Flow Rating _____
    Maximum Pressure Rating _____
    Port Size _____
    Number of Solenoids _____
    Solenoid Voltage _____

(4) Turn the valve body so that the mounting surface is visible, and label the diagram below to indicate location of pressure, tank, and working ports.

(5) Draw a complete schematic of the directional control valve using standard ISO symbols. Connect system lines to the neutral block of the DCV symbols.

(6) Follow the disassembly instructions as outlined in the manufacturer's service literature, disassemble the directional control valve.

(7) What is the condition of the spool surface?

(8) Referring to the service literature, list any missing parts including part numbers.

(9) Reassemble the directional control valve.
(10) For the valve you disassembled, draw three diagrams showing the spool and body in the three operation positions and the connections to a single rod and double acting cylinder. Use colors to trace high pressure and tank pressure through the valve to the actuator.

Lab Report

Part 2   Relief Valves

(1) Obtain a relief valve to disassemble.

(2) Complete the following list of specifications for the relief valve.

    Manufacturer: _____

    Model Number: _____

(3) Record the following preliminary observations making use of the catalog information provided in the laboratory.

    Flow Rate _____

    Pressure Adjustment Range _____

    Type of Main Port Connections (Pipe, Subplate, or Flange) _____

(4) Obtain the manufacturer's service information bulletin for the valve. Look over the information carefully before starting disassembly. Follow any special instructions while making the disassembly and reassembly. Do not attempt to remove any seats.

(5) List any missing parts according to the manual.

Part	Abnormal Condition

(6) Reassemble the valve into the same condition as it was at the beginning of the exercise.

(7) Does the valve have one or two springs on opposite side of the main spool movement? Explain why two springs might be used.

(8) Can this model valve be externally piloted and how? Use ISO symbols to draw the sketch.

(9) Explain the purpose of the small orifice in the spool.

(10) Is there a hole through the center of the spool? Regardless of your answer, what is the purpose of such a hole?

Lab Report 9 Analysis of Flow Control Valves

Name _____ Study ID _____

Test Stand _____ #. Date _____

Table R.9.1 Simulation Result of Throttle Valve

Throttle Valve DF1 Opening Scale: __1 mm__ .

System Pressure p_1/bar	System Loader Pressure p_2/bar	Target Pressure p_2/bar	Pressure Differential Δp/bar	Flow Rate $q/(\text{L} \cdot \text{min}^{-1})$
		15		
		20		
		25		
		30		
		40		

Table R.9.2 Simulation Result of Throttle Valve

Throttle Valve DF1 Opening Scale: __1.1 mm__ .

System Pressure p_1/bar	System Loader Pressure p_2/bar	Target Pressure p_2/bar	Pressure Differential Δp/bar	Flow Rate $q/(\text{L} \cdot \text{min}^{-1})$
		15		
		20		
		25		
		30		
		40		

Table R. 9. 3 Experiment Result of Throttle Valve

Throttle Valve DF1 Opening Scale: __scale 3__ .

System Pressure p_1/bar	System Loader Pressure p_2/bar	Target Pressure p_2/bar	Pressure Differential Δp/bar	Flow Rate $q/(\text{L} \cdot \text{min}^{-1})$
		15		
		20		
		25		
		30		
		40		

Table R. 9. 4 Experiment Result of Throttle Valve

Throttle Valve DF1 Opening Scale: __scale 4__ .

System Pressure p_1/bar	System Loader Pressure p_2/bar	Target Pressure p_2/bar	Pressure Differential Δp/bar	Flow Rate $q/(\text{L} \cdot \text{min}^{-1})$
		15		
		20		
		25		
		30		
		40		

Table R.9.5 Simulation Result of Speed-regulating Valve

Speed-regulating valve DF3 Opening Scale: __2 L/min__ .

System Pressure p_1/bar	System Loader Pressure p_2/bar	Target Pressure p_2/bar	Pressure Differential Δp/bar	Flow Rate $q/(L \cdot min^{-1})$
		15		
		20		
		25		
		30		
		40		

Table R.9.6 Simulation Result of Speed-regulating Valve

Speed-regulating Valve DF3 Opening Scale: __6 L/min__ .

System Pressure p_1/bar	System Loader Pressure p_2/bar	Target Pressure p_2/bar	Pressure Differential Δp/bar	Flow Rate $q/(L \cdot min^{-1})$
		15		
		20		
		25		
		30		
		40		

Table R. 9. 7 Experiment Result of Speed-regulating Valve

Speed-regulating Valve DF3 Opening Scale: __scale 4__ .

System Pressure p_1/bar	System Loader Pressure p_2/bar	Target Pressure p_2/bar	Pressure Differential Δp/bar	Flow Rate $q/(\text{L} \cdot \text{min}^{-1})$
		15		
		20		
		25		
		30		
		40		

Table R. 9. 8 Experiment Result of Speed-regulating Valve

Speed-regulating Valve DF3 Opening Scale: __scale 7__ .

System Pressure p_1/bar	System Loader Pressure p_2/bar	Target Pressure p_2/bar	Pressure Differential Δp/bar	Flow Rate $q/(\text{L} \cdot \text{min}^{-1})$
		15		
		20		
		25		
		30		
		40		

Questions

Compare the throttle valve with the speed-regulating valve, and draw the characteristic curves of the two kinds of valves in the following chart according to the experiment data. What is the difference between throttle valve and speed-regulating valve and illustrate the reasons? Draw different curves in different line styles.

Lab Report

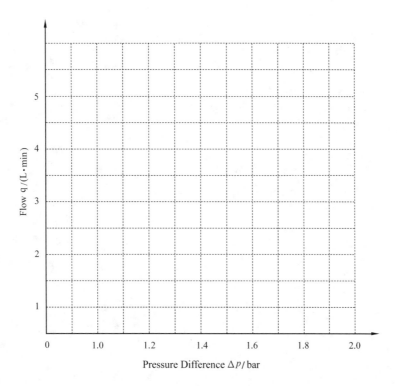

Figure R. 9. 1　Flow Characteristic Curve

Lab Report 10 Familiar with Hydraulic Circuits

Name _____   Study ID _____
Test Stand _____ #. Date _____

Figure R. 10. 1   Sequence Control Circuit　　**Figure R. 10. 2   Flow Control Circuit**
　　　　　　Components Lists　　　　　　　　　　　　　　**Components Lists**

Figure R. 10. 3　Pressure Control Circuit Components Lists   Figure R. 10. 4　Speed-shift Circuit Components Lists

Lab Report 11 Proportional Control System

Name _____   Study ID _____

Test Stand _____ #.   Date _____

Table R. 11. 1 Experiment Result without Load

Proportional Gain K_p	Position/ mm	Actual Valve Voltage/V	Error Voltage/V
1			
2			
5			
6			
8			
10			
12			

Table R. 11. 2 Experiment Result with Load

Load/bar	Position/mm	Actual Value Voltage/V
0		
5		
10		
15		
20		

(continued)

Load/bar	Position/mm	Actual Value Voltage/V
25		
30		
35		
40		

Questions

(1) What happens when proportional coefficient changes from 1 to 5? Is the response faster or slower?

(2) What is the difference between the ideal proportional valve and the actual proportional valve?

Lab Report 12 Hydraulic Throttle Control Experiment for Engineering Machinery

Name _____ Study ID _____
Test Stand _____ #. Date _____

(1) When only the hydraulic motor rotates clockwise, record the amount of the by-pass oil return and T-oil return.

Table R. 12. 1 By-pass Oil Return and T-oil Return

No.	By-pass Oil Return/(L · min^{-1})	T-oil Return/(L · min^{-1})
1		
2		
3		
4		
5		

(2) In separate operation, is the running speed of the actuator proportional or inversely proportional to the corresponding handle amount of deflection?

(3) In parallel operation 1, does the speed of the hydraulic motor increase or decrease? Why?

(4) When parallel operation is acting, does the oil flow first to low pressure load or high pressure load? Why?

Lab Report 13 Hydraulic Load Sensing Control Experiment for Engineering Machinery

Name _____   Study ID _____

Test Stand _____ #.   Date _____

(1) In single action, record the outlet pressure of the hydraulic pump and the pressure of the LS oil passage, and analyze the difference between the pressure of the LS oil passage and the outlet pressure of the pump.

Table R. 13. 1 Pressure of LS Oil Line and Pump Outlet Pressure in Single Actuator Operation

Hydraulic Cylinder Rising Process			Hydraulic Motor Clockwise Rotation Process		
p_{LS}/bar	p_P/bar	$(p_{LS}-p_P)$/bar	p_{LS}/bar	p_P/bar	$(p_{LS}-p_P)$/bar

(2) Explore the relationship between the speed of the hydraulic motor and the flow through the hydraulic cylinder in parallel operation 1.

Table R. 13. 2 The Relationship between Motor Speed and Hydraulic Cylinder Flow in Parallel Operation

$n/(\text{r} \cdot \text{min}^{-1})$	$q/(\text{L} \cdot \text{min}^{-1})$

(3) In parallel operation, when the output flow of hydraulic pump is **sufficient**, does the oil flow preferentially to the low pressure load or the high pressure load? At this time, will the operation of the actuators be affected by each other? Why?

(4) In parallel operation, when the output flow of hydraulic pump is **insufficient**, does the oil flow preferentially to the low pressure load or the high pressure load? At this time, is the working principle of the system equivalent to what kind of speed control circuit?

Lab Report 14 LUDV Control Experiment for Engineering Machinery

Name _____ Study ID _____

Test Stand _____ #. Date _____

(1) In single action, is the running speed of the actuator proportional or inversely to the corresponding handle amount of deflection? What is the maximum flow at each actuator?

(2) In parallel operation 1, record the flow value on the hydraulic cylinder and the hydraulic motor, and calculate the pressure difference across the throttle valve at this time.

(3) Explore the relationship between the speed of the hydraulic motor and the flow through the hydraulic cylinder in parallel operation 2, and what rules can you draw?

Table R.14.1　The Relationship between Hydraulic Motor Speed and Cylinder Flow

$n/(\text{r}\cdot\text{min}^{-1})$	$q/(\text{L}\cdot\text{min}^{-1})$

(4) Does the oil flow preferentially to the lower load or the higher load? Will the operation of the actuators be affected by each other at this time? Why?